Concepts in Chemistry

An Experimental Introduction to Reaction Kinetics

Concepts in Chemistry

Titles in this series:

An Introduction to Chromatography
Critical Readings in Chemistry
A Guide to Understanding Basic Organic Reactions
Orbitals and Chemical Bonding
An Introduction to Chemical Energetics
Spectroscopy in Chemistry
The Chemical Electron
Structure and Behaviour
Crystals and Crystal Structure
Organometallic Chemistry of the Main Group Elements

An Experimental Introduction to Reaction Kinetics

M. A. Atherton and J. K. Lawrence
Royal Grammar School, Newcastle upon Tyne

Longman

Longman Group Limited
London
Associated companies, branches and representatives throughout the world

First published 1970

ISBN 0 582 32145 X

Printed in Great Britain by
Spottiswoode, Ballantyne and Co Ltd
London and Colchester

Preface

There is considerable evidence that reaction kinetics is one of the most misunderstood topics in chemistry at school level. This problem seems to arise directly from the unqualified application of the law of mass action to the stoichiometric equations for reactions. Experimental aims such as 'to verify the law of mass action' which have appeared in some textbooks show that this approach is dogmatic rather than empirical. When the approach is based on establishing the rate expression for each reaction experimentally, the recognition of elementary and complex reactions (and hence the need for reaction mechanisms) arises naturally: there can be no confusion about the 'prediction' of rate expressions from stoichiometric equations. The purpose of this book is to provide such an experimental approach.

It seems best to refer to the law of mass action only from the historical standpoint. Even when the law is found to 'predict' successfully the experimental rate expression for a reaction, it cannot be assumed that the reaction is necessarily elementary. (The hydrogen-iodine reaction is an example of this.)

The first three chapters of this book are designed as an elementary introduction and cover the work necessary, for instance, for the Nuffield O-level examination. They could be used by pupils in their pre-O-level year, or by pupils beginning the subject in the first year of the sixth form. The remainder of the book takes the treatment of the subject up to and beyond A-level. Some of the later material may well be useful for first year university courses in physical chemistry.

We do not expect, of course, that any sixth-former should carry out all the experiments in the text, but we have aimed to provide a wide range of tested experiments from which a selection may be made. However, results have been given for many of the experiments, so that pupils can have practice in the interpretation of results even when they have not performed the experiment for themselves. In the case of the organic reactions discussed, the results provide experimental evidence on which a discussion of mechanism can be based; if organic reaction mechanisms are to be more widely taught in schools, it is important that at least some of the experimental evidence for them should be produced.

Integrated rate expressions are introduced at the beginning of Chapter 4. We have found that sixth-formers do not always appreciate their significance in the interpretation of experimental results, and a clearer understanding can be gained by using plots of concentration against time and drawing tangents to the curve. If examining boards require a knowledge of the integrated forms, these can be introduced towards the end of the course on kinetics. Both methods of interpreting experimental results are illustrated in the examples later in the book.

We should like to record our appreciation of the generous help which we have received. Professor P. G. Ashmore of the University of Manchester Institute of Science and Technology has provided invaluable help and guidance. He criticized most of the book in detail and also helped to sort out our ideas in the early stages; we have also found his articles and books on reaction kinetics lucidly written and most informative. Mr K. McDonald of the Royal Grammar School, Newcastle upon Tyne, read the first draft and pointed out many obscurities, errors and inconsistencies. Professor E. E. Aynsley of Newcastle University read the manuscript and gave us much early encouragement, and Dr P. Jones of the same university helped with the experiments on the 'flowing clock'. We are indebted to Mrs N. Atherton, who typed the manuscript, and to those other colleagues at the Royal Grammar School who offered constructive advice and criticism. All contributed much toward the production of this book, but any errors and inadequacies which remain are entirely our own responsibility.

Royal Grammar School
Newcastle upon Tyne

M. A. Atherton
J. K. Lawrence

Contents

Preface v

1 An introduction to rates of chemical reactions 1

2 The measurement of rates 5

3 Further investigations on rates—concentration and catalytic effects 16

4 The determination of rate expressions and orders of reaction 25

5 Other experimental techniques for determining reaction rates 55

6 Order and mechanism 68

7 Theories of reaction rates 82

8 More about catalysis 98

9 The hydrolysis of alkyl halides and esters 117

10 More aspects of the kinetics of gaseous reactions 136

11 Kinetics and equilibrium 149

Appendices 157

Index 166

1 An introduction to rates of chemical reactions

The thorough investigation of any chemical reaction involves finding answers to a number of questions including:

1 What are the substances which react?
2 What is produced?
3 What are the relative quantities of the reactants and the products involved in the reaction?

When the answers to these questions are known, an equation can be written.

There may be many difficulties in this part of the investigation. As an example, consider the action of heat on solid sodium bicarbonate. It is soon discovered that carbon dioxide and water vapour are formed. But are these products of the decomposition of sodium bicarbonate, or have they been formed by the reaction of the solid with some gas in the air? Is the sodium bicarbonate damp? At this stage, various *guesses* could be made about the equation for the reaction:

$$2NaHCO_{3(c)} = Na_2CO_{3(c)} + H_2O_{(g)} + CO_{2(g)}$$
$$NaHCO_{3(c)}(\text{damp}) = NaOH_{(c)} + CO_{2(g)} + xH_2O_{(g)}$$
$$2NaHCO_{3(c)} = Na_2O_{(c)} + 2CO_{2(g)} + H_2O_{(g)}$$

Only an investigation of the products of the reaction, followed by quantitative experiments, would show that the first equation is the one which applies under normal laboratory conditions. Even so, if the reaction conditions were changed sufficiently, the last two equations, and perhaps others, could become relevant. For instance, at a high enough temperature sodium carbonate decomposes.

$$Na_2CO_{3(c)} = Na_2O_{(c)} + CO_{2(g)}$$

The original investigation could be continued with questions such as:

4 How long does it take for the change to be completed?
5 Can the reaction be made to occur more quickly?

Such questions as these, concerning the rate (speed) of reactions and the conditions which affect the rate, form the basis of *Reaction Kinetics*. This involves many measurements of the rate of chemical change. Chemists use these experimental data to increase their understanding of how chemical reactions occur.

Reactions go at different speeds

Many familiar reactions occur very rapidly. In those between silver nitrate and sodium chloride solutions, and between barium chloride and sodium sulphate solutions, a white precipitate appears immediately.

$$Ag^+_{(aq)} + Cl^-_{(aq)} = AgCl_{(c)}$$
$$Ba^{2+}_{(aq)} + SO^{2-}_{4(aq)} = BaSO_{4(c)}\dagger$$

† The symbol 'aq' indicates that the ion is present in aqueous solution.

Not all precipitation reactions are quite so fast. When a solution of sodium thiosulphate is mixed with dilute hydrochloric acid the precipitate of sulphur takes several seconds to appear.

$$S_2O^{2-}_{3(aq)} + 2H^+_{(aq)} = H_2O_{(l)} + SO_{2(aq)} + S_{(c)}$$

Another reaction which occurs in solution, the neutralization of an acid with an alkali, is accompanied by the evolution of heat. The use of a thermometer as a stirring rod shows that all the heat is evolved during a very short period of time.

Some reactions between solids are also very rapid, even at times explosive. High temperature ignition of powdered mixtures such as aluminium powder and iron(III) oxide, and powdered magnesium and sulphur, produces explosions. It is interesting to note that these mixtures could be left indefinitely at room temperature without any detectable change occurring. The important point here is that, once begun, the reaction occurs rapidly.

In everyday life there are many examples of reactions which proceed very slowly: an iron girder may take decades to rust completely; the green patina which appears on bronze statues and copper water tanks exposed to the atmosphere forms slowly like rust.

Some of the slowest reactions are those involved in geological processes. The chemical weathering of rocks is caused by the action of water and atmospheric gases. For example, limestone is slowly dissolved by rainwater, a dilute solution of carbonic acid.

$$CaCO_{3(c)} + \text{'}H_2CO_3\text{'}_{(aq)} = Ca(HCO_3)_{2(aq)}$$

Again, the sandy streaks seen on granite mountains show that weathering has occurred. The slowness of the weathering processes may be realized by examining the stonework of old buildings, where the layer of crumbly weathered rock is rarely more than half an inch thick.

In between the extremes of very fast and very slow reactions are many which are of particular interest because accurate measurements of their rates are possible.

Factors which affect reaction rates

A consideration of some simple reactions will help to establish some of the factors controlling rate. Magnesium and water react so slowly at room temperature that effervescence is hardly noticeable. Several days are necessary to obtain sufficient gas for testing. However, if steam is passed over heated magnesium, the metal burns quickly in the steam and a test-tube of gas is obtained in a few seconds. Marble chips react with hydrochloric acid slowly enough for the reaction to be used in a Kipp's apparatus, but with an equal weight of powdered calcium carbonate the reaction would be much too rapid. When heated in air, steel wool glows red and burns slowly, but if it is then transferred to a gas jar of oxygen it burns brightly. The presence of suitable 'impurities' sometimes causes a marked change in the rate of a reaction. In the preparation of oxygen, the decomposition of potassium chlorate is accelerated by the addition of fairly small quantities of manganese dioxide.

In the above examples it appears that reaction rates may be affected by a variety of factors. The effects of changing some factors are shown in the following experiments. In each experiment only one factor is varied at a time. At the end of the five experiments, make a list of five factors which affect reaction rates.

Experiment 1.1 Reaction of metals with acids
Estimate the reaction rate by observing the vigour of effervescence.

1. Using a 20 mm length of magnesium ribbon as a standard size, cut out similar sized pieces of other metal foils such as aluminium, zinc, lead and copper. Place in different test-tubes and add equal volumes of dilute hydrochloric acid to each.

2. Compare the magnesium ribbon reaction in (1) with that of a roughly equal weight of magnesium powder with the same volume of dilute hydrochloric acid.

3. Into two test-tubes pour equal volumes of dilute hydrochloric acid, and to one add about 1 cm^3 of copper sulphate solution. Add a similar strip of zinc foil to each tube.

4. Into three test-tubes pour equal volumes of dilute hydrochloric acid and to each add two drops of copper sulphate solution. Heat one tube gently and one until the liquid nearly boils. Add a similar strip of zinc foil to each tube.

5. Half fill one test-tube with dilute hydrochloric acid and half fill another with concentrated hydrochloric acid which has been diluted with an equal volume of water. Add a similar strip of zinc foil to each tube.

The control of reaction rates

In industry, a process may need to be accelerated or slowed down. There is a need to speed up slow processes so that they become economically worthwhile, and some potentially explosive reactions have to be carefully controlled. Accelerating a process is usually done by increasing the working temperature, by using a catalyst, and sometimes by increasing the pressure. In the production of ammonia by the Haber Process, the working temperature is about 500°C, the pressure is about 200 atmospheres, and an iron catalyst is used. On the other hand, in atomic reactors, control rods are used to moderate the nuclear reactions so that they do not proceed too quickly.

There are many examples in everyday life where the control of reaction rate is important. The various methods of rust control are examples of this, and so is the use of sulphur dioxide in food preservation. Also the slowing of the decay of food at low temperatures is the basis of the frozen food industry. In an internal combustion engine a mixture of air and petrol vapour is compressed and exploded by a spark from the spark plugs. 'Knocking' (or 'pinking') occurs when the explosion happens spontaneously prior to sparking, so causing uneven combustion. The reaction rate is reduced by using an 'anti-knock' additive, like lead tetraethyl ($Pb(C_2H_5)_4$), which prevents the reactions which cause explosions during compression. Combustion then only takes place during sparking. Both petrol and diesel engines are good examples of how very rapid reactions can be controlled and made useful.

Processes which occur in nature are controlled in a variety of ways. For instance, the fermentation of sugar to alcohol is achieved by the addition of yeast. Again, bacterial action causes food decay, and also helps in the decay of vegetable matter to peat or garden compost. In the human body, enzymes in saliva and stomach juices assist digestion. The presence of chlorophyll in green leaves of plants assists in the important process called photosynthesis, but here light from the sun is also necessary. Carbon dioxide and water can combine under these conditions to produce carbohydrates.

$$6CO_{2(g)} + 6H_2O_{(l)} \xrightarrow[\text{chlorophyll}]{\text{light}} C_6H_{12}O_{6(aq)} + 6O_{2(g)}$$

Questions and topics for discussion

1 Cork does not react with oxygen in the air at room temperature, but cork dust suspended in air has been known to explode.

2 Reactions where a cation and an anion react to form a precipitate appear to be instantaneous. Can reactions be instantaneous? Is there a reason why such reactions should be so fast?

3 How might the rate of combustion in a candle flame be increased?

4 The use of a pressure cooker reduces the time taken for vegetables to cook.

5 Rock salt, put on icy roads during winter, accelerates the rusting of motor vehicles.

6 A mixture of hydrogen and oxygen may be left indefinitely without appearing to react, but if a spark is passed, or heat is applied, it may explode.

7 Damp hay in stacks and oily rags stored in cupboards may ignite spontaneously.

2 The measurement of rates

What is meant by the rate of a chemical reaction?

To answer this question it will be useful to see how the rates of various everyday processes are measured. The rate at which water flows from a tap may be expressed by the quantity of water flowing from the tap in a certain time. This could be measured by collecting water for, say, 1 minute and finding either its volume or weight. The flow rate could then be expressed in gallons, litres, pounds or grammes per minute. Similarly, in a cricket match the scoring rate would be found from the number of runs scored in a given time, and may be expressed as the number of runs per hour. Speeds of vehicles are also rates and, of course, these are measured as distance travelled in a certain time.

There are two very important consequences which follow:

1 Each rate is expressed as a quantity (like volume, weight or length) achieved in a given time. Units of rate, therefore, are quantities per unit time.
2 Each rate is the average rate for the length of time chosen. Within this time interval the rate may have varied considerably. For instance, in the tap water example, the water pressure may have changed during the minute the water was being collected.

The second point could be shown by collecting the water in a large measuring cylinder and noting the volume of water in the cylinder every 10 seconds. Suppose that over the minute during which the water was being collected the water pressure decreased uniformly. The results might look like this:

Volume of water cm^3	*Time taken* s
0	0
50	10
90	20
122	30
148	40
169	50
186	60

From these figures the average rates over each 10-second interval can be calculated.

Time interval s	*Average rate* cm^3/s
0–10	50/10 or 5·0
10–20	4·0
20–30	3·2
30–40	2·6
40–50	2·1
50–60	1·7

However, the average rate over the whole minute is 186/60 or 3·1 cm^3/s which is not the same as the rate over any of the 10-second intervals.

Even the rates over these smaller time intervals are only average rates, and it is possible that the rate varied during the 10-second intervals as well. However, as the time interval is shortened the possibility of considerable variation becomes much less, until over a small enough interval the rate can be considered constant. This gives rise to such expressions as: 'the rate *at* time 10 seconds' (i.e. at that instant when the reaction has been going for 10 seconds).

How can rates at given times be found? If a graph is plotted using the above figures, a smooth curve can be drawn going very close to, or through, all the points

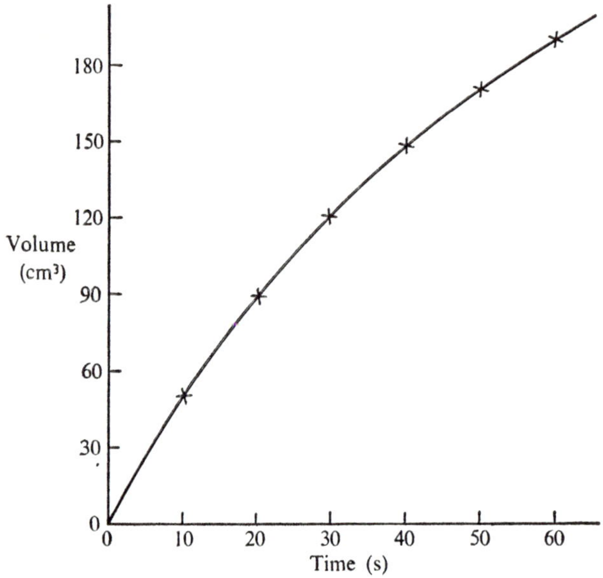

Fig. 2.1. Variation of water volume with time

(see Fig. 2.1). The curve represents the *continuous* change in water volume with time.

A closer examination of the curve for the 20 to 30 second interval shows that the average rate (122–90/30–20) in this interval is the slope of the chord AC ($\tan.\widehat{ACB} = AB/BC$)—see Fig. 2.2.

The average rate during the time interval 24 to 26 seconds could be estimated by drawing the necessary chord and calculating its slope. As the time interval becomes smaller the chord becomes shorter until, at *one* specified time, the chord becomes the tangent at the point where A and C coincide. This is shown at time 25 seconds where the slope of the tangent DE is equal to the flow rate at time 25 seconds. In Fig. 2.2 it will be noted that chord AC and tangent DE have nearly the same slope. This is not always the case, however, but over a *small* time interval the slope of the chord is a good approximation to the slope of the tangent at the time in the middle of the time interval concerned.

The idea of a rate in chemistry is basically the same as for any everyday example such as those discussed above. The reaction rate must be expressed as a quantity divided by time. Moreover, the rate frequently varies during the reaction. Again, it would be possible to quote an average rate for any interval of time using a table of experimental results, but only a graph could show the continuous change of rate as the reaction proceeds.

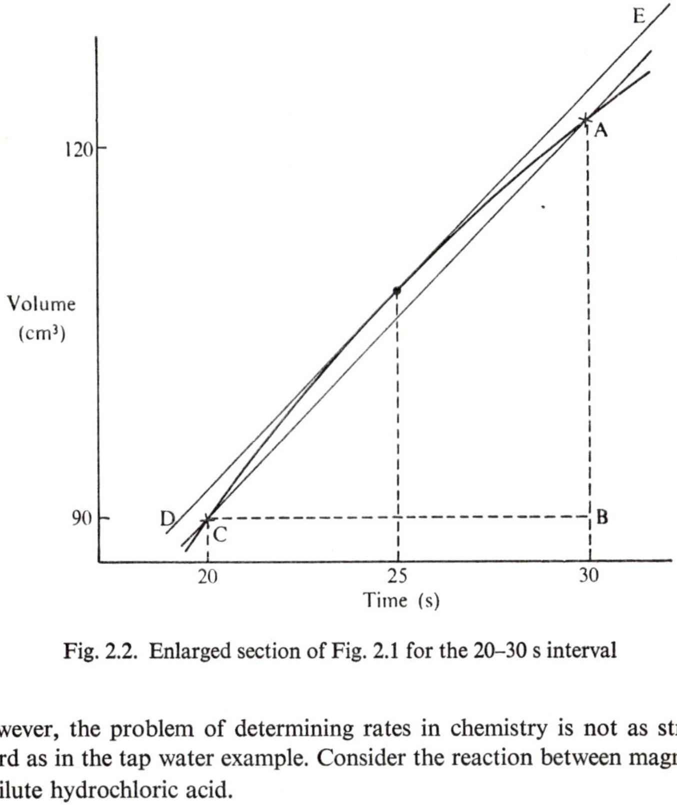

Fig. 2.2. Enlarged section of Fig. 2.1 for the 20–30 s interval

However, the problem of determining rates in chemistry is not as straightforward as in the tap water example. Consider the reaction between magnesium and dilute hydrochloric acid.

$$Mg_{(c)} + 2H^{+}_{(aq)} = Mg^{2+}_{(aq)} + H_{2(g)}$$

Instead of having to consider simply the quantity of water, there are four different substances whose quantities vary as the reaction proceeds. The quantities of the two reactants, magnesium and the acid, decrease, and the quantities of the two products, magnesium chloride and hydrogen, increase. Is it necessary to measure the changes in the quantities of all four substances at various times during the reaction? Even if it were possible it certainly would be a tedious task.

Fortunately, it is only necessary to measure the change in quantity of one of the four substances. This comes about because, if the rate of change of one is known, then calculations based on the equation will indicate what the other rates are. In the above equation,

1 mol Mg reacts with 2 mol H^+ to produce 1 mol each of Mg^{2+} and H_2†

If it was found by experiment that 1 mol of hydrogen was produced in 1 hour, then it follows that the average rates over that hour would be:

Rate of dissolving of magnesium = 1 mol $hour^{-1}$.
Rate of decrease of acid concentration = 2 mol $hour^{-1}$ (for a given volume of solution).
Rate of evolution of hydrogen = 1 mol $hour^{-1}$.
Rate of production of magnesium ions = 1 mol $hour^{-1}$ (for a given volume of solution).

It must be emphasized that the equation itself gives no information about how fast a chemical reaction goes, but merely shows the proportions in which the reactants react and the products form. However, if one rate has been determined experimentally, then the rates of change for the other reactants and products are indicated by the equation.

In expressing a reaction rate, the following points should be considered:

1 Always state the units. Convenient units for gases are $cm^3\ s^{-1}$ or $cm^3\ min^{-1}$, but these may be converted to mol s^{-1} or mol min^{-1} by calculation. For solutions, it is usual to use concentrations in mol dm^{-3}, and the rate units then become mol $dm^{-3}\ s^{-1}$ or mol $dm^{-3}\ min^{-1}$.
2 A sign convention is useful. If the rate is expressed in terms of reactants used up, a minus sign is used, e.g. $-R(Mg)$, where R represents rate, and the substance whose changes in quantity are being considered is placed in brackets after R. If the rate is expressed in terms of products formed, a plus sign indicates that there is a gain in quantity instead of a loss, e.g. $+R(H_2)$.
3 It should be stated clearly whether a quoted reaction rate is an average rate or a rate at a given time.

Before an experiment is attempted, it must be decided which reactant or product is to be used to assess the rate of reaction. In the magnesium–dilute acid example there are four possibilities because there are four substances involved in the reaction. The changes in the concentration of acid and the concentration of magnesium ions could not be followed visually, so the dissolving of the magnesium or the evolution of hydrogen would be a better choice.

The simplest way to investigate the reaction between magnesium and dilute hydrochloric acid is to measure the time taken for pieces of magnesium to dissolve. Such measurements only give the average rate over the whole of the reaction: they do not show the rates at intermediate stages. If, however, the volume of hydrogen is measured at several times from the start of the reaction, it is possible to assess the rates at different stages of the reaction. These two methods will be dealt with in turn.

† The term 'mol' will be used in place of such terms as 'g. molecule', 'g. formula' and 'g. ion'. 'Molarity', expressed by the symbol 'M', will represent a concentration in mol dm^{-3}.

Experiment 2.1 Measurement of time taken for magnesium to dissolve

There are several factors which might affect the time taken for the magnesium to dissolve: (*a*) concentration of acid; (*b*) temperature; (*c*) the form of the magnesium. The normal procedure when there are three variables is to maintain two of them constant and to find the effect of varying the third. In each case the experiment consists of adding a piece of magnesium to 50 cm^3 of acid and measuring the time taken for the magnesium to dissolve. The starting quantities are given in Table 2.1.

Table 2.1

	Temperature	*Magnesium*	*Hydrochloric acid*
Exp. 2.1(*a*) Concentration of acid varied	Room temperature	30 mm strips of clean ribbon	x cm^3 2M HCl + $(50-x)$ cm^3 water, where $x = 50$, 40, 30, 20
Exp. 2.1(*b*) Temperature varied	Heat the diluted acid to the required temperature before adding the Mg. Temperatures: 25°, 35°, 45°, 55°C	30 mm strips of clean ribbon	30 cm^3 2M HCl + 20 cm^3 water
Exp. 2.1(*c*) Form of Mg varied	Room temperature	(i) 0·03 g powder (ii) 30 mm clean ribbon (iii) 9 mm clean wire. These all have approximately the same weight	50 cm^3 2M HCl

(A convenient way of ensuring that all forms of magnesium have the same weight is to use a 30 mm piece of ribbon as the standard weight on one pan of a chemical balance. The amount of wire and powder which have the same weight as the ribbon can be found by counterbalancing the ribbon with an equal weight of wire or powder on the other pan of the balance.)

The average rate over the whole reaction in each case will be equal to the weight of magnesium used, divided by the time taken for it to dissolve. Since the same weight is used in each reaction the average rate is inversely proportional to the time taken.

The results of these experiments show that temperature, the concentration of the acid, and the form of the magnesium are all important in controlling the rate of reaction. Try to suggest reasons why these factors have this effect.

These are other factors which could be varied: the quantity of magnesium; the shape of the container; and other acids and metals could be used. If any of these are of interest, devise experiments to see whether these factors affect the reaction rate.

Measurement of the volume of hydrogen evolved as the reaction proceeds

There are several devices for measuring varying volumes of gas, but the two simplest are syringes and gas burettes. The latter are either obtained from manufacturers of apparatus or may be made from two tapless burettes as shown in Fig. 2.3.†

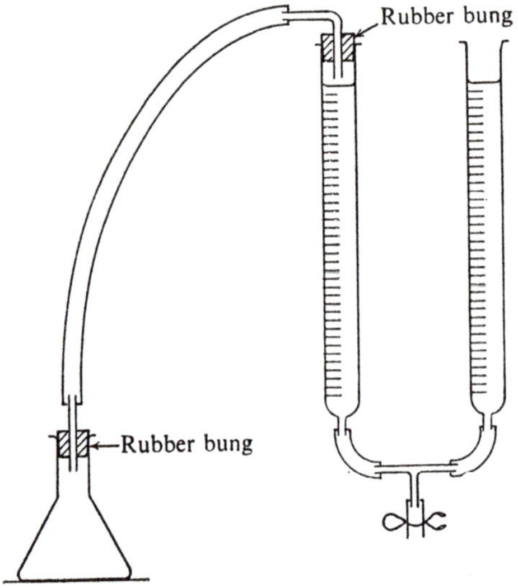

Fig. 2.3. A simple gas burette

Experiment 2.2 How does the volume of hydrogen vary during the reaction between magnesium and hydrochloric acid?

Pour 100 cm^3 of 0·50 M hydrochloric acid into the flask, slide in 15 mm of clean magnesium wire and quickly insert the bung. If a gas burette is used, allow water to run out so that the water surfaces in the tubes are at the same level, read the volume on the left-hand burette and either note the time or start the stop-clock. This is the reading on the burette at time $t = 0$ and should be taken as soon as possible after the reaction has started. The volume reading at $t = 0$

† Purchased gas burettes usually have a capacity of 200 cm^3. If the 50 cm^3 gas burettes are used, some of the reactions cannot be followed for the times stated, but the experiments still give useful results.

does not have to be zero, provided this is remembered when the graph is plotted. Repeat the readings at 1-minute intervals for about half an hour, remembering each time to make the water surfaces in the tubes level.

Discussion of results

Study the table of results. How does the volume of hydrogen evolved during each minute vary as the reaction proceeds? What change is taking place in the rate of reaction throughout the 30 minutes? Is it possible to explain this change by referring to changes in the reaction mixture?

Plot a graph of volume in cm^3 of hydrogen along the vertical axis against time in minutes along the horizontal axis. The curve should pass through the origin since at $t = 0$ no hydrogen has been evolved. Why does the curve begin steeply and gradually flatten? What would have happened to the curve if the experiment

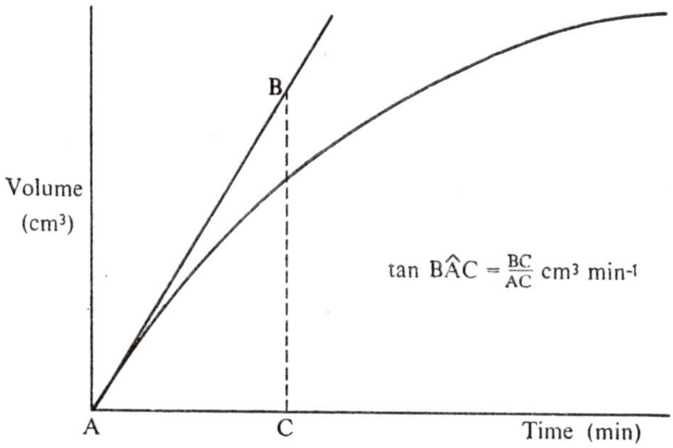

Fig. 2.4. The initial rate tangent

had been continued until all the magnesium had dissolved? Is it possible to predict from the graph the time at which the reaction would have stopped and what the volume of hydrogen would then have been? Why is it difficult to get this information from such a curve?

Make an estimate of the volume of hydrogen which should be evolved by 15 mm (0·05 g) of magnesium wire under normal laboratory conditions. Use the fact that the G.M.V. of any gas is about 24 dm^3 at room temperature and pressure. Is this volume of the same order of magnitude as the volume estimated from the graph?

Divide the curve into 5-minute intervals and calculate from the graph the volumes of hydrogen evolved in each interval. From these figures, express the average rate of reaction in $cm^3\ min^{-1}$ for each interval. Each of these average rates is approximately equal to the rate at the time mid-way through the 5-minute period.

One rate of particular importance in the experiments in this chapter is the *initial rate*. This is the rate at $t = 0$. From the graph it should be clear that this is the highest rate achieved in the reaction. It can be found by calculating the slope of the tangent to the curve at the origin as in Fig. 2.4.

Experiments 2.3 to 2.6 use measurements of the volume of hydrogen evolved in reactions between metals and dilute hydrochloric acid to investigate in more detail the effects of the factors varied in Experiment 2.1. A further factor, the addition of an impurity, is also dealt with. Again, in each experiment one factor is varied whilst the others are kept constant.

Experiment 2.3 The effect of acid concentration on rate

Reagents: 100 cm^3 0·25 M hydrochloric acid
15 mm clean magnesium wire.

Procedure

Measure the volume of hydrogen at 1-minute intervals for about 10–15 minutes.

Plot the results and those from Experiment 2.2 on the same graph.

Discussion of results

What differences are there in the shapes of the two curves? Compare the total volume of hydrogen evolved after 12 minutes in this experiment with that evolved after the same time in Experiment 2.2. If the reactions were allowed to proceed until all of the magnesium had dissolved, would the same volume of hydrogen be liberated in each case?

Is it easy to estimate the time taken for the reaction in 2.3 to stop? Would it be longer or shorter than the time required in Experiment 2.2?

Calculate the initial rate of reaction. This should be approximately one half of the initial rate in 2.2. What might be the significance of this?

Experiment 2.4 The effect of the form of the magnesium on rate of reaction

(*a*) 0·05 g (50 mm) clean magnesium ribbon;
100 cm^3 0·5 M hydrochloric acid.

Take readings every half-minute for about 15 minutes.

(*b*) 0·05 g magnesium powder;
100 cm^3 0·5 M hydrochloric acid.

Try to take readings every quarter-minute for the first minute and then every half-minute until the reaction stops.

This experiment is difficult to perform. Place the magnesium powder in a small, wide test-tube suspended by a thread in the hydrochloric acid. In this way the volume reading at $t = 0$ can be taken before the reaction starts. The reaction is started by tipping the reaction flask sideways, and the flask should be held in this position until the reaction is finished.

Discussion of results

Plot the results for Experiments 2.4(*a*) and (*b*) on the same graph and compare them with the graph from Experiment 2.2.

Why are the shapes of the curves so different? At what points on the graphs for 2.4(*a*) and (*b*) is it clear that the reactions have stopped? What are the volumes of hydrogen evolved at the end of the reactions and how do these compare with the estimate made in the discussion of the results of Experiment 2.2? Compare the initial rates in 2.4(*a*) and (*b*) with that in 2.2.

What can be concluded about the effect of the state of subdivision of the magnesium on the rate? Why should this factor have such a dramatic effect?

Experiment 2.5 The effect of temperature on rate of reaction

(*a*) Reagents as for Experiment 2.2.

Immerse the flask containing the acid in a large beaker of water. Heat the water to 41°C and turn off the Bunsen burner. Add the magnesium wire and proceed as in previous experiments, reading the volume of hydrogen at 1-minute intervals until the temperature of the water has fallen to 39°C. If a large volume of water is used, this temperature drop will occur slowly, and the reaction can be followed for several minutes. The average temperature for the reaction will be 40°C.

(*b*) The procedure is the same as in Experiment 2.5(*a*) but the flask is immersed in crushed ice. Read the volume of hydrogen at 1-minute intervals for about 15 minutes.

Discussion of results

Plot the results for 2.5(*a*) and (*b*) on the same graph and compare this with the results from Experiment 2.2. Calculate initial rates in each case.

What conclusions can be reached about the effect of temperature on the rate of this reaction?

Experiment 2.6 The effect of copper sulphate on the rate of reaction between zinc and hydrochloric acid

(*a*) *Reagents:* A piece of granulated zinc weighing 0·30 g;
Hydrochloric acid: 75 cm^3 water and 25 cm^3 concentrated hydrochloric acid.

Proceed as in Experiment 2.2, measuring the volume of hydrogen at 1-minute intervals for about 15 minutes. Observe carefully during the initial stages of the reaction.

(*b*) Proceed as in Experiment 2.6(*a*) but add about 2 cm^3 of approximately 0·5 M copper sulphate solution to the acid before starting the reaction.

Read the volume of hydrogen at 1-minute intervals for about 15 minutes. Plot the results for 2.6(*a*) and (*b*) on the same graph and calculate the slopes where the curves are steepest.

It is most important in each of these reactions to use *one* piece of zinc, and the two pieces should be of the same shape and have similar lustres.

Discussion of results

The shapes of both curves are different from those for the magnesium-acid reactions (see Appendix A). What is the main difference? Explain this in terms of the change of rate with time.

Is it possible to measure initial rates for these reactions from the graphs? Why are these not 'true' initial rates as defined in the discussion of results for Experiment 2.2? The reluctance of zinc to liberate hydrogen is surprising in view of its position in the reactivity series for metals. Try to suggest an explanation for this. What effect does the copper sulphate have? Is it very marked? The appearance of red copper metal around the zinc shows that the zinc has displaced the copper from solution.

$$Zn_{(c)} + Cu^{2+}_{(aq)} = Zn^{2+}_{(aq)} + Cu_{(c)}$$

The presence of both copper and zinc may have some bearing on the problem of why copper sulphate affects the reaction rate.

It should be clear from these experiments that there are at least four important factors which affect the rates of reactions between metals and dilute hydrochloric acid:

1. The concentration of the acid;
2. The state of subdivision of the metal;
3. The temperature;
4. The presence of other substances.

To these can be added a fifth, the nature of the reactants. In Experiment 2.6(*a*) the use of zinc instead of magnesium decreased the rate of reaction considerably, so that a much higher acid concentration was required to give a convenient rate.

Further practical exercises

1. Using the procedure in Experiment 2.2, but replacing the 0·50 M hydrochloric acid with 100 cm^3 of 0·50 M acetic acid, plot a curve of volume of hydrogen in cm^3 against time in minutes.

Compare the curve with that in Experiment 2.2 and try to explain the result. N.B. The reaction can *still* be represented by the equation

$$Mg_{(c)} + 2H^{+}_{(aq)} = Mg^{2+}_{(aq)} + H_{2(g)}$$

2. Investigate the rate of reaction between calcium carbonate and 2 M (bench) hydrochloric acid by a gravimetric method.

Place 20 g of marble chips in a conical flask and completely cover with 40 cm^3 of 2 M hydrochloric acid. Loosely plug the neck of the flask with cotton wool to avoid loss of acid spray. Read the weight on a direct reading balance each minute and plot a graph.

In this reaction, the only gaseous substance is carbon dioxide.

$$CaCO_{3(c)} + 2H^{+}_{(aq)} = Ca^{2+}_{(aq)} + H_2O_{(l)} + CO_{2(g)}$$

The loss in weight of the flask plus contents over each minute therefore represents the weight of carbon dioxide evolved in that minute.

Extend the experiment by varying (one at a time) (*a*) the acid concentration (use the same *volume* as above), and (*b*) the state of subdivision of the calcium carbonate (use 20 g again).

Can the same conclusions be drawn as in the reactions between metals and dilute hydrochloric acid?

3. Investigate the reaction between magnesium and bench nitric acid. Use the procedure in Experiment 2.1(*b*) to determine the effect of temperature on the rate of reaction. (Temperatures up to about 70°C can be used.)

Then determine the effect on the reaction rate of adding one drop of copper(II) sulphate solution at several different temperatures. Repeat the experiment using solutions of the following salts: cobalt(II) nitrate, chromium(III) sulphate and zinc sulphate.

Using the same graph paper, plot suitable graphs to summarize all the results. Are there any general conclusions which can be drawn?

4. Investigate the effect of temperature on the rate of the reaction between potassium iodide and nitric acid.

Add 10 cm^3 of concentrated nitric acid to 50 cm^3 of bench nitric acid and divide the resulting solution into three parts. Keep one portion at room temperature, heat another at 50°C, and boil the third. Add half a test-tube of concentrated potassium iodide solution to each acid portion and note the time taken for the iodine colour to appear. Care should be taken in the handling of concentrated nitric acid.

3 Further investigations on rates—concentration and catalytic effects

A. Rates and concentration

In the investigation of the reaction between magnesium and hydrochloric acid in Chapter 2, it was found that the rate depended on the acid concentration. Experiment 2.2 showed that during the reaction the rate decreased as the acid concentration decreased. A mathematical expression of this might possibly take the form $+R(H_2) \propto [H^+]^n$, where square brackets denote concentration in mol dm^{-3}, and n is a whole number or fraction with a positive value. This experiment, together with Experiment 2.3, suggested that halving the acid concentration halved the initial rate. If this is so, then

$$+R(H_2) \propto [H^+] \qquad \text{and } n = 1$$

Further investigations would be needed to establish this proportionality more firmly.

This chapter is concerned partly with an attempt to establish some relationships between rates of reaction and concentrations of reactants. The two reactions to be studied are the reaction between sodium thiosulphate and hydrochloric acid and the decomposition of hydrogen peroxide. If the reaction rates are proportional to $[\text{reactant}]^n$ then it will be possible to determine values for n. It should not be expected that such a relationship will hold in all cases and more complicated expressions may apply.

The reaction between sodium thiosulphate and hydrochloric acid

Experiment 3.1
Shake a few crystals of sodium thiosulphate with water in a test-tube until a solution is formed, and then add dilute hydrochloric acid.

The sulphur produced is white initially because it is very finely divided. After a few minutes the characteristic yellow colour appears.

$$S_2O_{3(aq)}^{2-} + 2H_{(aq)}^{+} = H_2O_{(l)} + SO_{2(aq)} + S_{(c)}$$

This is a precipitation reaction involving ions and such reactions usually appear instantaneous. Why is this one exceptional?

Determination of the rate of reaction

The obvious change in this reaction is the precipitation of sulphur. This can be used to estimate the rate of reaction in the following way. The solutions of thiosulphate and acid are mixed in a beaker which is then placed on a cross marked on a piece of paper. The cross is viewed from above through the solution, and the time taken for it to disappear is measured.

If the rate of precipitation of sulphur is large, the time taken will be small, whereas, if the rate is small, the time will be large. Provided that the total volume of the solutions is kept constant, then the depth in the beaker is constant. This ensures that the amount of sulphur needed to obscure the cross is the same in different experiments. Under these conditions, $+R(\mathrm{S}) \propto 1/t$, where $+\mathrm{R(S)}$ is the rate of sulphur precipitation and t is the time taken for the cross to disappear.

The rate determined in this way is the average rate over the fraction of reaction followed. In Experiments 3.2 and 3.3 a small fraction of the reaction has been completed when the cross disappears so that the average rate is approximately the initial rate.

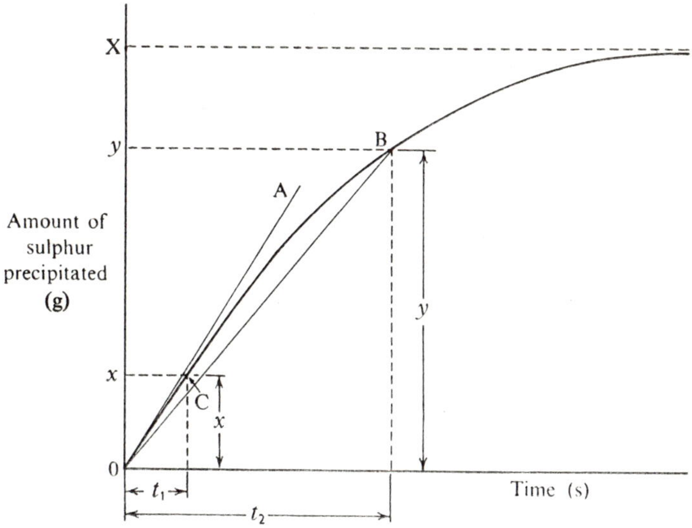

Fig. 3.1. Possible variation of amount of precipitated sulphur with time

In Fig. 3.1 it can be seen that the initial rate is expressed to a reasonable degree of accuracy by the average rate between time 0 and time t_1 seconds.

Average rate between time 0 and time $t_1 = \dfrac{x}{t_1}$ g s^{-1}

(i.e. slope of chord OC)

Therefore, slope of tangent at time 0 $\sim$ slope of chord OC $= \dfrac{x}{t_1}$ g s^{-1}, since the slopes of OA and OC are closely similar.

Therefore, initial rate (slope of tangent OA) $\sim \dfrac{x}{t_1}$ g s^{-1}.

If x is the same in every experiment where the initial rate is being estimated, then:

Initial rate $\propto \dfrac{1}{t_1}$ s^{-1}.

From the graph, x/X (where X is the total amount of sulphur precipitated) *must* represent a small fraction of the total reaction. If it were not a small fraction, the slope of tangent OA would not approximate to the initial rate, as can be seen from the chord OB, drawn for a large fraction, y/X, of the reaction. Its slope is y/t_2, which is very different from the initial rate of x/t_1. Ideally, measurements should be made on the first few per cent of the reaction.

Since there are two reactants in this reaction, it is necessary to find how the initial rate of sulphur precipitation depends on the concentration of each. The procedure is to keep one concentration constant whilst the other is varied.

Experiment 3.2 Variation of the initial rate with the concentration of thiosulphate ions

Reagents

Each group will require about 500 cm³ of approximately 0·1 M sodium thiosulphate solution to which a pinch of sodium carbonate has been added (this prevents premature precipitation of sulphur) and 120 cm³ of concentrated hydrochloric acid.

Procedure

Into one beaker dispense 20 cm³ of acid from a 25 cm³ measuring cylinder. Measure the volumes of sodium thiosulphate solution and water in a 100 cm³ measuring cylinder and then pour them into a second beaker.

Mix the two solutions by pouring twice from one beaker to another. Start the stop-clock the moment the two solutions begin to mix. Note the time, t, taken for the cross on the piece of paper to disappear.

Volume $Na_2S_2O_3$ cm³	*Volume* H_2O cm³	*Volume* HCl cm³	t s	$1/t$ s^{-1}
100	0	20		
90	10	20		
80	20	20		
70	30	20		
60	40	20		
50	50	20		

Plot a graph of $1/t$ against volume of sodium thiosulphate solution. The latter is a measure of the concentration of thiosulphate ions.

Discussion of results

It should be possible to draw a straight line through the points on the graph. Decide whether this line passes through the origin. If there is direct proportionality between $+R(S)$ and the thiosulphate concentration, then this should go through the origin. When the thiosulphate concentration is zero, there is no reaction and t must equal infinity, so making $1/t$ equal to zero as well. If the line does not pass through the origin, decide whether or not this is due to experimental error.

Experiment 3.3 Variation of initial rate with the acid concentration

Reagents

Each group will require 600 cm³ of approximately 0·4 M sodium thiosulphate solution and 40 cm³ of bench (2 M) hydrochloric acid diluted to 200 cm³ with water.

Procedure

The technique is the same as in Experiment 3.2. Use a 25 cm^3 (or 50 cm^3) measuring cylinder to measure the acid and water.

Quantities:

Volume $Na_2S_2O_3$ cm^3	*Volume* H_2O cm^3	*Volume* HCl (0·4 M) cm^3	t s	$1/t$ s^{-1}
100	0	60		
100	10	50		
100	20	40		
100	30	30		
100	40	20		

Plot a graph of $1/t$ against volume of acid. Since this volume is proportional to the hydrogen ion concentration, the graph shows the relationship between $+R(S)$ and this concentration.

Discussion of results

In Fig. 3.2 some typical results have been plotted. There is no question of this plot being a straight line. At low acid concentrations, $+R(S)$ is approximately

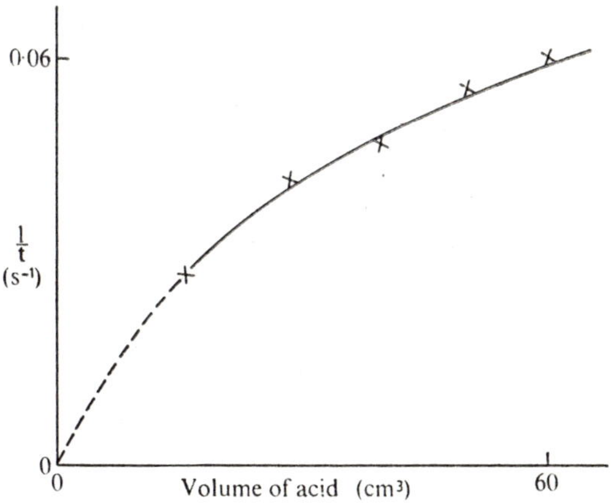

Fig. 3.2. Variation of the initial rate with the concentration of acid in Experiment 3.3

proportional to the acid concentration because the line is more or less straight. At higher acid concentrations, $+R(S)$ increases more and more slowly with increasing concentration. Given that the acid concentration begins to slow down the increase in $+R(S)$ at higher concentrations, try to think of some possible relationships between $+R(S)$ and this concentration.† Question 2 in the exercises at the end of the chapter investigates this problem further.

† See Appendix A.

The decomposition of hydrogen peroxide

$2H_2O_{2(aq)} = 2H_2O_{(l)} + O_{2(g)}$

In this reaction the interest lies in how $-R(H_2O_2)$ varies with the peroxide concentration. This relationship *may* be of the form $-R(H_2O_2) \propto [H_2O_2]^n$, in which case the problem will be to determine *n*.

A convenient way of measuring the reaction rate is to follow the increase in volume of oxygen evolved with time. Hence it will be $+R(O_2)$ that is measured. From the balanced equation, $+2R(O_2) = -R(H_2O_2)$, if both rates are expressed in the same units. Thus $+R(O_2)$ will also be proportional to $[H_2O_2]^n$, if the relationship between rate and concentration is of this form.

The experimental results will be volumes (V_t cm^3) of oxygen at given times (*t* min), and yet the calculations will require a knowledge of the peroxide concentration. Using gas volume calculations, each V_t could be converted to mol of oxygen and hence to mol of hydrogen peroxide. This would be tedious, however, and there is an easier way of expressing this concentration.

Let V_∞ be the final volume of oxygen when the reaction is complete.

$V_\infty \propto [H_2O_2]_0$, where $[H_2O_2]_0$ is the initial concentration of hydrogen peroxide. (All of this has decomposed when V_∞ is measured.)

After *t* min let V_t cm^3 of oxygen have been evolved.

Then $V_t \propto [H_2O_2]_t$, where $[H_2O_2]_t$ is the concentration of hydrogen peroxide used up in the time *t*.

$(V_\infty - V_t) \propto ([H_2O_2]_0 - [H_2O_2]_t) \propto [H_2O_2]$, where $[H_2O_2]$ is the concentration of hydrogen peroxide *remaining* in the reaction mixture at time *t*. Therefore, if $+R(O_2) \propto [H_2O_2]^n$, then $+R(O_2) \propto (V_\infty - V_t)^n$.

The decomposition is slow normally, but it can be accelerated by the addition of a small quantity of manganese(IV) oxide.

Experiment 3.4

The experiment can be performed using one of the three pieces of apparatus described in Experiment 2.1. The quantities below give a V_∞ of about 100 cm^3.

Procedure

Dilute 5·0 cm^3 of '20 volume' hydrogen peroxide solution with 45 cm^3 of water. Make this solution either neutral or slightly alkaline by the addition of dilute caustic soda solution. (One drop of 2 M solution is usually sufficient.) This ensures that none of the catalyst dissolves in the hydrogen peroxide.

Tip 0·050 g of manganese(IV) oxide† into the hydrogen peroxide solution and start the stop-clock immediately. Equalize the water levels in the gas burette and read the volume. Shake the flask vigorously at first, so as to mix the reagents thoroughly, and swirl consistently for the rest of the experiment. Keep the flask immersed in cold water to maintain a steady temperature. (The reaction is exothermic.)

Read the volume every minute until between 80 and 90 per cent of the oxygen has been evolved. Leave the apparatus set up until the reaction has stopped and then take the V_∞ reading.

Plot a graph of V_t against *t*.

† The manganese(IV) oxide is the finely powdered, precipitated solid of purity not less than 70 per cent (obtainable from B.D.H.).

Discussion of results

The results should give a graph like that in Fig. 3.3.

The rates at given times can be estimated by drawing tangents at convenient intervals. In Fig. 3.3 the slope of tangent AB is the rate at time = 0, volume = 0; the slope of tangent CD is the rate when time = t_1, and volume = v_1; and the slope of tangent EF is the rate when time = t_2, and volume = v_2.

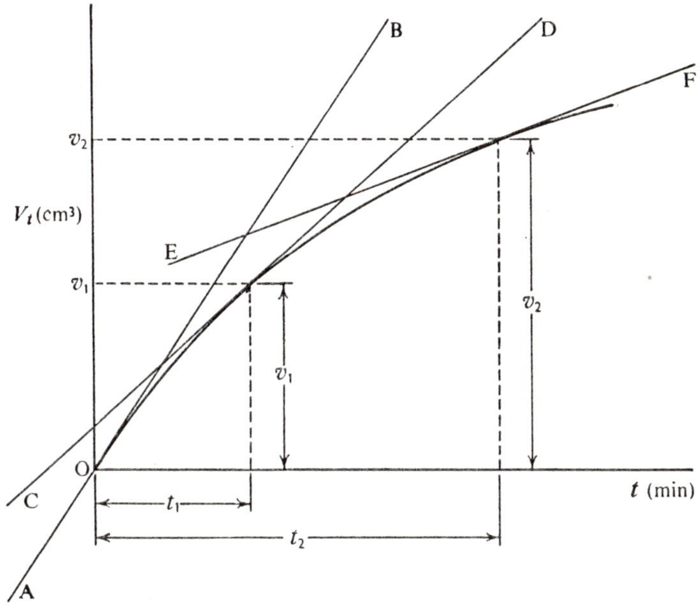

Fig. 3.3. Graph of V_t against t (showing some tangents)

Carefully draw tangents to the curve at about the following times: 0, 2, 4, 6, 9, 12, 16, and 20 min. Construct a table of results with these headings:

Slope of tangent	*time*	V_t	$(V_\infty - V_t)$	$(V_\infty - V_t)^2$
$cm^3\ min^{-1}$	t min	cm^3	cm^3	cm^6

Plot graphs of tangent slope against $(V_\infty - V_t)$ and $(V_\infty - V_t)^2$ on the same graph paper. If the first plot is a straight line then $+R(O_2) \propto [H_2O_2]$, but if the second is a straight line then $+R(O_2) \propto [H_2O_2]^2$. Try to decide whether $+R(O_2)$ is more likely to be proportional to $[H_2O_2]$ or to $[H_2O_2]^2$.

Some of the error involved in drawing tangents can be avoided in the following way. A smooth curve is drawn through the points and a rectangular two-sided mirror (e.g. a stainless steel mirror) is placed with its plane at right angles to the graph paper and approximately along the normal to the curve at the point where the tangent is required. The mirror and curve are viewed from above and the mirror is tilted slightly and then rotated until the reflection of the curve on one side of the mirror appears continuous with the line on the other side of the mirror.

By tilting the mirror the opposite way, the procedure can be repeated. The mirror edge is now along the normal to the curve, and when the normal has been drawn, the tangent, which is perpendicular to the normal, can be constructed. Alternatively, the relationship

$$\text{slope of tangent} = -\frac{1}{\text{slope of normal}}$$

can be used.

One further graph is possible. If $+R(O_2) \propto [H_2O_2]^n$, then tangent slope $\propto [H_2O_2]^n$ or tangent slope $= k\,(V_\infty - V_t)^n$, where k is a constant.

Taking logarithms,

$$\log_{10}(\text{tangent slope}) = \log_{10} k + n \log_{10}(V_\infty - V_t)$$

If a graph of $\log_{10}$(tangent slope) against $\log_{10}(V_\infty - V_t)$ is a straight line then its slope will be n.

Plot this graph and decide whether a reasonable straight line can be drawn. If it can, its slope is likely to be 1, suggesting that $+R(O_2)$ is proportional to $[H_2O_2]$.

Conclusions

Experiments 3.2, 3.3 and 3.4 all show that the relationships between rates and concentration are not necessarily simple. They also show that there is no obvious way of predicting such relationships from any simple theory. Each proportionality must be investigated by carefully designed experiments.

The experimental results themselves are often difficult to interpret, and it is important not to make hurried conclusions. It is often necessary to obtain results over a very wide range of conditions, and to plot them in several different ways, before any definite pattern emerges. In view of this, it should be noted that much more work would be required on the above experiments before it would be possible to establish their kinetics thoroughly.

B. Rates and catalysis

It has been mentioned already that the presence of substances other than reactants can speed up a reaction considerably. A few drops of copper sulphate solution increase the rate of evolution of hydrogen from zinc and hydrochloric acid. Also, the rate of decomposition of hydrogen peroxide was found to be accelerated by the addition of a small amount of magnanese(IV) oxide. Such 'impurities' are known as *catalysts*, and the process is known as *catalysis*.

The following experiments will help to establish some of the features of catalytic action.

Experiment 3.5 Catalysis of the decomposition of hydrogen peroxide

(*a*) *Is the manganese(IV) oxide used up in the reaction?*

Weigh, as accurately as possible, about 1 g of manganese(IV) oxide and add it to about 100 cm³ '20 vol' hydrogen peroxide, which has been made just alkaline by the addition of sodium hydroxide solution. When the reaction has stopped, filter the mixture, and dry the manganese(IV) oxide and filter paper in an oven. Weigh the solid and filter paper and subtract the weight of the filter paper. Decide whether there has been any significant loss of weight of the catalyst during the reaction.

(*b*) *Does the form of the catalyst have any effect on the reaction rate?*
Perform the reaction as in Experiment 3.4 but only follow it for about 5 minutes. Use 0·05 g of granular manganese(IV) oxide. Draw a graph of the results and work out the initial rate. Compare this with the initial rate in Experiment 3.4 where 0·05 g of finely powdered solid was used.

(*c*) *Does the form of the catalyst change during the reaction?*
Allow the reaction in (*b*) to go to completion. Carefully observe the size and condition of the granules before and after the experiment. Compare this change with any change occurring in the appearance of the granules after they have been immersed in water for the same length of time.

(*d*) *Can other substances catalyse the decomposition?*
Add small amounts of the following substances to separate test-tubes containing a few cm^3 of '20 vol' hydrogen peroxide. Note whether or not effervescence begins.

1. Lead(IV) oxide.
2. Copper(II) oxide.
3. Blood.
4. Sand.
5. A few drops of iron(III) chloride solution followed by sufficient bench caustic soda to precipitate iron(III) hydroxide. After a short time, add a little dilute acid to dissolve the precipitate. As a control experiment, repeat the procedure of adding about the same quantities of alkali and acid to hydrogen peroxide in the absence of iron(III) chloride.
6. Add a few drops of concentrated ammonia solution to a small quantity of cobalt(II) nitrate solution until a brown solution is obtained. Add two drops of the resulting solution to some '20 vol' hydrogen peroxide. As in (5), perform a control experiment in which no cobalt(II) nitrate is used.

Discussion of results

In (*d*), only small amounts of catalyst are required to produce some effervescence, and, in the case of manganese(IV) oxide and lead(IV) oxide, the effect is very marked. This is a common feature of many catalysed reactions. Try to think of more examples which illustrate this point.

In many catalysed reactions the amount or concentration of the catalyst has some effect on the reaction rate. In the decomposition of hydrogen peroxide it appears that the amount of manganese(IV) oxide used has a great influence on the reaction rate. Question 3 in the exercises at the end of this chapter suggests an investigation of this.

In (*a*) it will have been noted that the catalyst was not used up in the reaction, although in (*c*) the granules did become powdery on their surfaces. The amount of surface area exposed to the hydrogen peroxide is very important, since finely powdered manganese(IV) oxide gives a much higher initial rate than the granular solid. In (*d*), several other powdered materials with large surface areas catalyse the decomposition. In the case of iron(III) hydroxide, the formation of the precipitate substantially increased the rate of effervescence, and its solution by dilute acid caused the effervescence almost to stop.

However, the relative effects of these surfaces varied considerably so that the chemical nature of the surface may be important as well as the amount of surface. It may be significant, for instance, that many metallic oxides have a marked catalytic effect, whereas the effect of sand is small under these conditions.

Not only solids catalyse the decomposition of hydrogen peroxide. In blood, there is a biological catalyst (enzyme) which causes some increase in effervescence. A very marked effect is also seen when an ammoniacal solution of cobalt(II) nitrate is added.

Practical exercises

1. Using the quantities in Experiment 3.2 and using the same experimental technique, find a way of obtaining plots of $1/t$ against volume of thiosulphate solution at about 0°C and at 60°C. Compare these plots with the one in Experiment 3.2, and comment on the results.
2. Make further investigations on the relationship between $+R(\mathrm{S})$ and the acid concentration as in Experiment 3.3. Use 100 cm^3 of 0·1 M sodium thiosulphate solution in each experiment, but vary the volume of concentrated hydrochloric acid from 10 cm^3 to 50 cm^3, diluting this to 50 cm^3 with water in each case.
3. Investigate the effect of varying the amount of manganese(IV) oxide on the initial rate of the decomposition of hydrogen peroxide. Add half, and then double, the amount of catalyst used in Experiment 3.4.
4. Investigate the effect of varying the initial concentration of hydrogen peroxide on the initial rate. Follow the procedure of Experiment 3.4 but use 5·0 cm^3 of '10 vol' and then 5·0 cm^3 of '5 vol' hydrogen peroxide.

   Do the results confirm, or disagree with, the conclusions in Experiment 3.4 concerning the value of n in $+R(O_2) \propto [H_2O_2]^n$?
5. Investigate the effect of temperature on the rate of decomposition of hydrogen peroxide. Proceed as in Experiment 3.4 but immerse the flask in a bath of ice.

4 The determination of rate expressions and orders of reaction

A. Introduction

Chapters 2 and 3 have been concerned with measuring rates of reactions and showing the effects of various factors on those rates. In some examples, simple relationships were found between the rate of reaction and the concentration of reactant. In this chapter, several reactions will be investigated with the aim of finding the detailed relationships between rate of reaction and concentration of reactants.

Consider the irreversible reaction between substances A and B to produce substances C and D. The rate of this reaction may be investigated by measuring

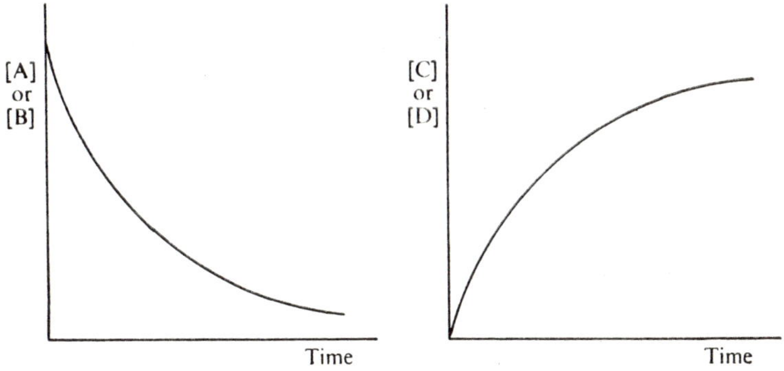

Fig. 4.1. Variation of reactant and product concentrations with time

the concentration of A, B, C, or D at several times during the reaction. Graphs of concentration against time are almost always of the type shown in Fig. 4.1. The rate of reaction for a certain concentration of reactant is numerically equal to the gradient of the curve at the point corresponding to that concentration.

Probably the largest source of error in finding the relationship between rate of reaction and concentration lies in the drawing of the tangents required to give the rate of reaction. One way in which such errors may be avoided will be illustrated for two simple relationships which may occur in practice:

1. The rate of reaction is proportional to the concentration of reactant. $-R(\text{A}) \propto [\text{A}]$.
2. The rate of reaction is proportional to the product of the concentrations of A and B. $-R(\text{A}) \propto [\text{A}][\text{B}]$.

1. $-R(A) \propto [A]$

If the concentration of A is c at time t, then using calculus notation,

$$-\frac{dc}{dt} \propto c$$

and

$$-\frac{dc}{dt} = k_1 c \dagger \tag{1}$$

The negative sign is needed because c decreases as t increases. This k_1 is the constant of proportionality. Integration of equation (1) gives

$$-\log_e c = k_1 t + p \tag{2}$$

where p is a constant which may be evaluated by substituting in equation (2) the boundary conditions $t = 0$, and the initial concentration of A $= c_0$. Hence

$$-\log_e c_0 = p$$

and equation (2) may be rewritten

$$\log_e c = -k_1 t + \log_e c_0 \tag{3}$$

Equation (3) has the form $y = mx +$ constant.

Thus if $-R(\mathrm{A}) \propto [\mathrm{A}]$, a graph of $\log_e c$ against time will be a straight line of slope $-k_1$ with the intercept on the $\log_e c$ axis equal to $\log_e c_0$.

There are other ways of writing equation (3):

$$c = c_0 \mathrm{e}^{-k_1 t} \tag{4}$$

$$\log_e \frac{c_0}{c} = k_1 t \tag{5}$$

or, in terms of logarithms to base ten,

$$+\log_{10} c = \frac{-k_1}{2{\cdot}30} t + \log_{10} c_0 \tag{6}$$

To test if $-R(\mathrm{A}) \propto [\mathrm{A}]$ represents the variation of rate of reaction with concentration, analyses for the concentration of A are made at a number of noted times during the reaction. These results may be plotted according to equation (6). If $-R(\mathrm{A}) \propto [\mathrm{A}]$ satisfactorily describes the variation of rate with concentration, the points will lie on a straight line. A curve shows that $-R(\mathrm{A}) \propto [\mathrm{A}]$ is not in agreement with the experimental results.

Alternatively, values of k_1 may be calculated by substituting the experimental data into equation (6) which would be rewritten.

$$k_1 = \frac{2{\cdot}30}{t} \log_{10} \frac{c_0}{c} \tag{7}$$

Constant values for k_1 show that $-R(\mathrm{A}) \propto [\mathrm{A}]$ is satisfactory, whereas, if the calculated values of k_1 show either a systematic increase or decrease as the reaction proceeds, an expression other than $-R(\mathrm{A}) \propto [\mathrm{A}]$ is required.

Those reactions for which equations (1) to (6) hold are said to be *first order* reactions. The constant k_1 is called the *rate constant* or *velocity constant* and, by its definition, [equation (1)], it is independent of the concentration of the

† See Appendix E.

reactant. It is, however, dependent on temperature, so that the temperature at which k_1 is measured should be quoted. Its units can be found from equation (1). Since dc/dt has the units concentration time^{-1}, k_1 must have the units time^{-1}.

Equation (1), which expresses the dependence of the rate of reaction upon the concentration of reactant, is called a *rate law*, *rate equation* or *rate expression.*

2. $-R(A) \propto [A][B]$

In this example, there are two concentrations which need to be evaluated. The easiest case is that in which the initial concentrations of A and B are equal, and A and B react in the molar proportion of 1:1. The concentrations of A and B then remain equal throughout the reaction, since the reactants are consumed at the same rate.†

Thus, for the reaction A + B = products, if $-R(\text{A}) \propto [\text{A}][\text{B}]$ and the initial concentrations of A and B are both equal, the rate expression is

$$-\frac{dc}{dt} \propto c^2 \tag{8}$$

If the rate constant is k_2,

$$-\frac{dc}{dt} = k_2 c^2 \tag{9}$$

Integration of this equation yields

$$\frac{1}{c} = k_2 t + q$$

where q is the constant of integration. Substituting $c = c_0$ when $t = 0$ gives

$$\frac{1}{c_0} = q$$

so that

$$\frac{1}{c} = k_2 t + \frac{1}{c_0} \tag{10}$$

and a graph of $1/c$ against t will be a straight line with its gradient equal to k_2 and the intercept on the $1/c$ axis equal to $1/c_0$. The applicability of equation (10) may be tested by substituting experimental values for c and t.

Reactions to which equations (9) and (10) apply are called *second order* reactions. The units of the rate constant are concentration^{-1} time^{-1} [See equation (9)]. Therefore, unlike the first order rate constant, k_1, the units of k_2 contain a concentration term. However, the value of k_2 is again independent of the concentration of A or B, but dependent on temperature.

† Had A and B reacted in the molar proportion of 1:2, then the rate of decrease in the concentration of B would have been double that for A.

$$-\frac{d[\text{A}]}{dt} = -\frac{1}{2}\frac{d[\text{B}]}{dt}$$

If a graph of $1/c$ against time is a straight line for the reaction in which A and B react in the molar proportion of 1:1, this is consistent, not only with

$$-\frac{d[\mathrm{A}]}{dt} = k_2[\mathrm{A}][\mathrm{B}] \tag{11}$$

but also with

$$-\frac{d[\mathrm{A}]}{dt} = k_2[\mathrm{A}]^2 \tag{12}$$

and

$$-\frac{d[\mathrm{A}]}{dt} = k_2[\mathrm{B}]^2 \tag{13}$$

In all three cases the total order of the reaction is two. If rate expression (12) applies, the rate of reaction is independent of the concentration of B. The reaction would then be *zero order* with respect to B and second order with respect to A (i.e. $-d[\mathrm{A}]/dt = k_2[\mathrm{A}]^2[\mathrm{B}]^0$). For the rate expression (13) the reaction is zero order with respect to A and second order with respect to B.

Experiments in which the initial concentrations of A and B, $[\mathrm{A}]_0$ and $[\mathrm{B}]_0$, are not equal would be necessary to show which of the three rate expressions is the relevant one. Integration of equation (11) in this case gives

$$\log_{10}\frac{[\mathrm{B}]}{[\mathrm{A}]} = \log_{10}\frac{[\mathrm{B}]_0}{[\mathrm{A}]_0} + ([\mathrm{B}]_0 - [\mathrm{A}]_0)\frac{k_2}{2{\cdot}30}.t$$

where [A] and [B] are the concentrations at time t. A plot of $\log_{10}[\mathrm{B}]/[\mathrm{A}]$ against t will be a straight line if the rate of reaction can be expressed by equation (11).

A special case arises when the concentration of one reagent, A, is so much larger than the concentration of B, that changes in the concentration of A during the reaction are negligible when compared with the changes in the concentration of B.

The rate expressions (11), (12) and (13) then become:

$$-\frac{d[\mathrm{A}]}{dt} = k_2'[\mathrm{B}],\dagger$$

$$-\frac{d[\mathrm{A}]}{dt} = k_2''$$

and

$$-\frac{d[\mathrm{A}]}{dt} = k_2[\mathrm{B}]^2$$

where $k_2' = k_2[\mathrm{A}]_0$ and $k_2'' = k_2[\mathrm{A}]_0^2$. The total orders are then 1, 0 and 2 respectively.

† A reaction which can give this rate expression is sometimes called a 'pseudo first order reaction' because under less special conditions it shows second order kinetics. However, the term 'pseudo' should be avoided by giving the experimental conditions when the order is quoted. Since the order of a reaction is an experimental quantity, it cannot be 'pseudo'; it merely changes with changes in the experimental conditions.

Reactions are known in which the total order is three or even four. What would be the units of the rate constants in each of these cases?

For the general rate expression

$$-\frac{d[A]}{dt} \propto [A]^m[B]^n \tag{14}$$

the reaction is mth order with respect to A, nth order with respect to B, and its total order is $(m + n)$. Experiments show that the values of n and m may be positive or negative integers, frequently –1, 0, 1, 2, and they may also be fractional. It is also possible that an expression such as (14) is incapable of satisfying experimental results; the simple concept of order is then inapplicable.

When the rate expression for a reaction is being investigated, it is necessary to study at least 50 per cent of the reaction and, if possible, even more, otherwise

Table 4.1. Some standard integrated equations and plots

Possible orders with respect to A	B	*Initial conditions*	*Rate expressions*	*Integrated rate expressions*	*Simplest function to plot against time t*
1. Reactants A and B → Products (A and B react in molar proportions 1:1)					
0	0	$[A]_0$	$-\frac{d[A]}{dt} = -\frac{d[B]}{dt} = k_0$	$[A] = [A]_0 - k_0 t$	$[A]$
Total order = 0					
1	0	$[A]_0 = [B]_0$	$-\frac{d[A]}{dt} = -\frac{d[B]}{dt} = k_1[A]$	$\log_e[A] = \log_e[A]_0 - k_1 t$	$\log_e[A]$
Total order = 1					
1	1	$[A]_0 = [B]_0$	$-\frac{d[A]}{dt} = -\frac{d[B]}{dt} = k_2[A][B] = k_2[A]^2$	$\frac{1}{[A]} = \frac{1}{[A]_0} + k_2 t$	$\frac{1}{[A]}$
Total order = 2					
1	1	$[A]_0 \neq [B]_0$	$-\frac{d[A]}{dt} = -\frac{d[B]}{dt} = k_2[A][B]$	$\log_e\frac{[B]}{[A]} = \log_e\frac{[B]_0}{[A]_0} + ([B]_0 - [A]_0)k_2' t$	$\log_e\frac{[B]}{[A]}$
Total order = 2					
2. Reactants A and B → Products (A and B react in molar proportions 1:2)					
1	1	$[B]_0 = 2[A]_0$	$-\frac{d[A]}{dt} = -\frac{1}{2}\frac{d[B]}{dt} = k_2[A][B] = 2k_2[A]^2$	$\frac{1}{[A]} = \frac{1}{[A]_0} + 2k_2 t$	$\frac{1}{[A]}$
Total order = 2					
1	2	$[B]_0 = 2[A]_0$	$-\frac{d[A]}{dt} = -\frac{1}{2}\frac{d[B]}{dt} = k_3[A][B]^2 = 4k_3[A]^3$	$\frac{1}{[A]^2} = \frac{1}{[A]_0^2} + 4k_3 t$	$\frac{1}{[A]^2}$
Total order = 3					

N.B. For reaction 2, other orders are also possible (e.g. 2 with respect to B, 0 with respect to A, and so on). Apart from the rateconstant in the integrated rate expression being preceded by integers other than 1, these possibilities are covered in reaction 1.

the order may not be apparent. In such experiments, properties of the system other than concentration may be used, provided the property can be related to the concentration of one of the reactants or products. For example, if a gas is produced during the reaction its changes of volume can be used as a measure of the extent of reaction at different times. This has been done already in experiment 3.4. Other properties which can be used are changes of colour, pH, optical activity and so on (see Chapter 5).

Half-life of reactions

The half-life, $t_{1/2}$, of a reaction is the time taken for one half of the reactant originally present to be consumed.

Expressions for the half-lives of reactions may be derived from the relevant integrated rate expressions. For a first order reaction

$$\log_e c = -k_1 t + \log_e c_0 \tag{3}$$

by substituting $c = \frac{1}{2}c_0$ and rearranging, the expression for $t_{1/2}$ is obtained.

$$t_{1/2} = \frac{\log_e 2}{k_1} \tag{15}$$

For a second order reaction

$$\frac{1}{c} = k_2 t + \frac{1}{c_0} \tag{10}$$

and by using the same substitution

$$\frac{2}{c_0} = k_2 t_{1/2} + \frac{1}{c_0}$$

or

$$t_{1/2} = \frac{1}{c_0 k_2} \tag{16}$$

It is possible to obtain expressions for the half-lives of reactions of other orders in a similar way. For a first order reaction

$$t_{1/2} \propto \frac{1}{c_0^0}$$

for a second order reaction

$$t_{1/2} \propto \frac{1}{c_0^1}$$

and for a reaction of order n

$$t_{1/2} \propto \frac{1}{c_0^{n-1}}$$

Equation (16) shows that the half-life of a second order reaction depends on the initial concentration of the reactant and the rate constant. As the initial concentration of the reactant decreases, the half-life increases. However, for a first order reaction [equation (15)], half-life is independent of the initial concentration of the reactant and depends only on the rate constant. Hence the time taken for c_0 to decrease by $c_0/2$ is equal to the time required for $c_0/2$ to decrease by $c_0/4$. Similar results follow for other fractional times. The times required for the reactant concentration to decrease from 1·0 mol dm^{-3} to 0·9 mol dm^{-3} is $t_{1/10}$ and is equal, for example, to the time required for the concentration to decrease from 0·30 mol dm^{-3} to 0·27 mol dm^{-3}.

Radioactive decay

The radioactive decay of protoactinium 234 can be studied using the method given in The Nuffield Chemistry Sample Scheme, Part II. In such an experiment the number of counts per second (measured with a rate meter and a liquid counter) was recorded at 5-second intervals and the results are shown in Fig. 4.2, where the

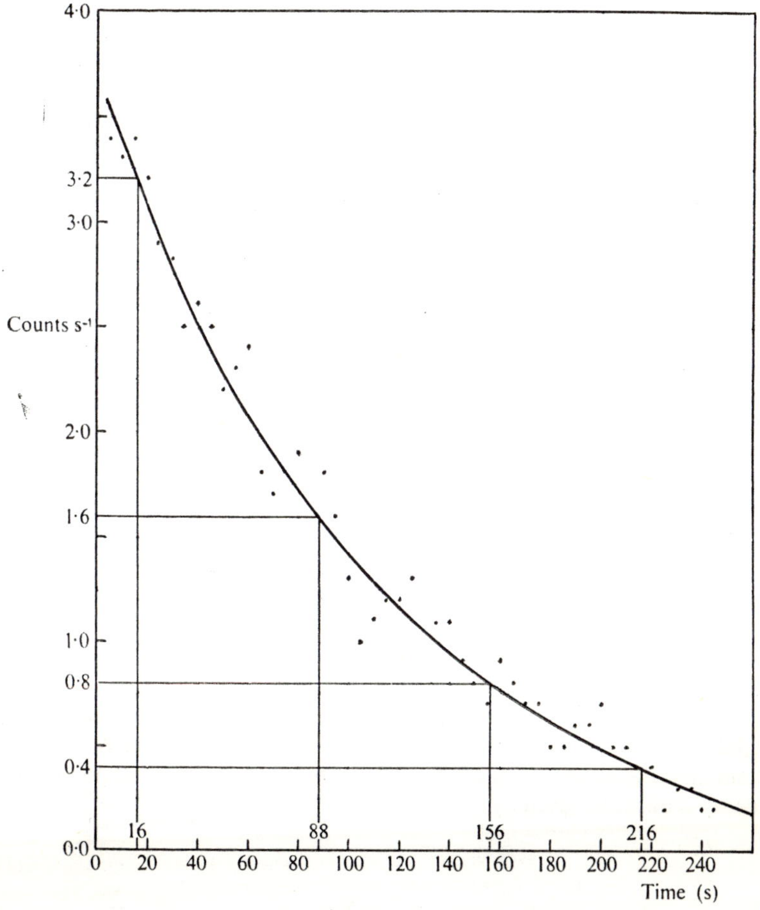

Fig. 4.2. Radioactive decay of protoactinium 234

number of counts per second is plotted against time. This count is proportional to the number of atoms disintegrating per second.

Inspection of this curve shows that the half-life is approximately constant at about 60–70 seconds. The decay thus appears to be a first order change. This conclusion may be verified from the graph of $\log_{10}$ (counts s^{-1}) against time in Fig. 4.3.

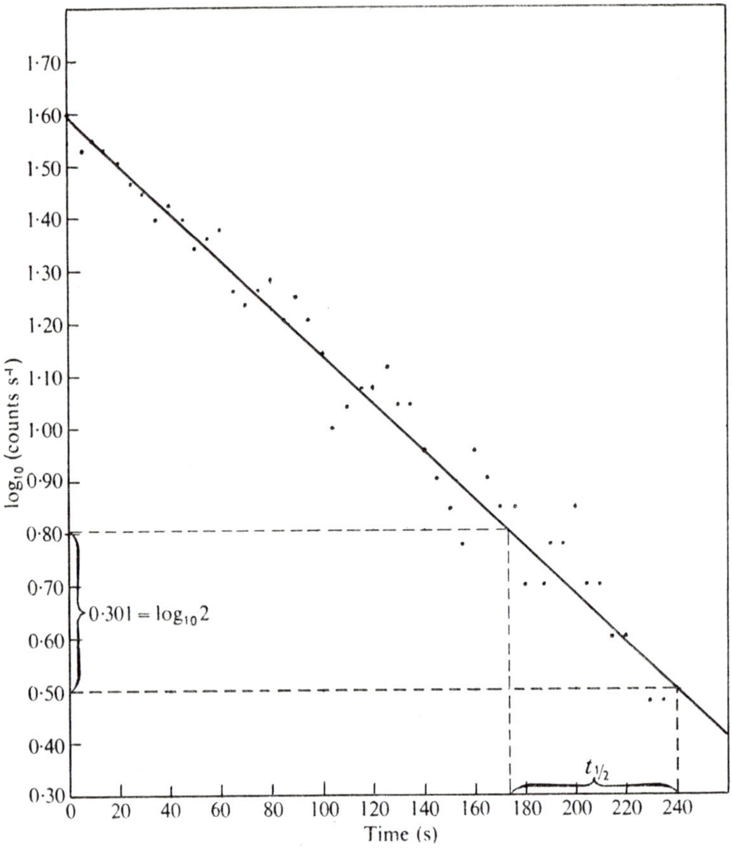

Fig. 4.3. A log–time plot of the decay of protoactinium 234

Results such as these can be interpreted statistically. Each atom of a radioactive element is subject to the chance λ that it will disintegrate within the next second. Thus for a sample of a radioactive element which contains N atoms at a certain time, the number of disintegrations in the next second will be λN. The rate of decay is therefore given by

$$-\frac{dN}{dt} = \lambda N \tag{17}$$

The constant λ, which is usually called the decay constant, has the units seconds^{-1}.

Equation (17) is similar to the first order rate expression (1), so that integration of this equation, followed by rearrangement, yields the following relationships:

$$\log_{10} N = -\frac{\lambda}{2{\cdot}30}t + \log_{10} \lambda N_0 \qquad (18)$$

$$N_t = N_0 \, e^{-\lambda t} \qquad (19)$$

$$t_{1/2} = \frac{2{\cdot}30}{\lambda} \log_{10} 2 \qquad (20)$$

The gradient of the line in Fig. 4.3 is $-4{\cdot}55 \times 10^{-3}\ s^{-1}$. From equation (18)

$$-\frac{\lambda}{2{\cdot}30} = -4{\cdot}5_5 \times 10^{-3}$$

Therefore

$$\lambda = 2{\cdot}3 \times 4{\cdot}5_5 \times 10^{-3} = 1{\cdot}0_5 \times 10^{-2}\ s^{-1}$$

and from equation (20)

$$t_{1/2} = \frac{2{\cdot}30 \times \log_{10} 2}{1{\cdot}05 \times 10^{-2}} = 66\ s$$

The half-life may also be determined from Fig. 4.3 by calculating the time required for $\log_{10} c$ to decrease to $\log_{10} c/2$ i.e. to $\log_{10} c - \log_{10} 2$. This is shown in Fig. 4.3, the time interval being $240 - 174 = 66$ s.

The decay constant λ is thus analogous to a first order rate constant. However, radioactive decay is very different from a first order chemical reaction. Radioactivity involves changes in the nucleus whereas any chemical change involves rearrangement of the valence electrons. Moreover, first order rate constants depend on the temperature at which the experiment is performed, whereas decay constants do not depend on temperatures which can be attained on earth.

B. Experimental work

Experiment 4.1 The decomposition of benzene diazonium chloride

The addition of sodium nitrite solution to a cooled, acidified aqueous solution of aniline produces a diazonium salt.

$$C_6H_5NH^+_{3(aq)} + HNO_{2(aq)} = C_6H_5N^+_{2(aq)} + 2H_2O_{(l)}$$

Diazonium salts are unstable and decompose in aqueous solution to produce phenol.

$$C_6H_5N^+_{2(aq)} + H_2O_{(l)} = C_6H_5OH_{(aq)} + N_{2(g)} + H^+_{(aq)}$$

For each mol of diazonium salt which decomposes, 1 mol of nitrogen is produced, and the rate of decomposition may be obtained by measuring the rate of formation of nitrogen.

Preparation of benzene diazonium chloride

Dissolve 1·0 cm³ of aniline in a mixture of 2·5 cm³ concentrated hydrochloric acid and 5·0 cm³ water. Cool the solution under running water and add, a few drops at a time, a solution of 0·80 g sodium nitrite in 2·0 cm³ water. The temperature of the mixture should be kept between 5 and 10°C and the mixture should be shaken after each addition. Continue adding the sodium nitrite solution until a drop of the reaction mixture gives an immediate blue colour with starch-potassium iodide paper. (Nearly all the sodium nitrite solution will be needed.)

This reaction mixture, when decomposed, usually gives about 250 cm³ of nitrogen. The amount of it used should be chosen according to the volume of the gas burette or glass syringe available.

Procedure

Place the benzene diazonium chloride solution with granules of pumice in a test-tube and connect this to the gas burette or syringe. (If pumice is not added, the nitrogen is not evolved steadily.) Completely immerse the test-tube in a beaker of water which has been heated to the required temperature. Allow the reaction mixture to attain the temperature of the water bath and then note the volumes of nitrogen, V_t, at known times t. It is not necessary that V_t should be zero at the start of the experiment. Maintain the temperature of the water bath at the chosen value.

The duration of the experiment and the frequency of readings depend on the temperature. At 45°C readings at about 1-minute intervals are required for at least 20 minutes. The apparatus needs to be left until the reaction has stopped, in order to measure the volume of nitrogen when decomposition is complete.

The experiment may be repeated at different temperatures in the range 30°C to 55°C.

(*a*) *Calculation of order from the rate of reaction at different times*

Let V_∞ be the volume of nitrogen produced when decomposition is complete and V_t the volume of nitrogen at time t. The amount of benzene diazonium

Table 4.2. The decomposition of benzene diazonium chloride at 45°C

Time min	*Time* s	V_t cm³	$V_\infty - V_t$ cm³	*Gradient* cm³ min⁻¹
3	30	19·5	178	7·1
6	0	36·5	161	6·5
7	45	47·5	150	—
10	0	60·0	138	5·65
12	0	71·5	126	—
14	0	83·0	115	4·8
16	0	91·5	106	—
18	0	99·5	98·0	4·15
22	0	115·5	82·0	3·5
24	0	121·5	76·0	—
∞		197·5		

chloride which *has* decomposed at time t is proportional to V_t, and the initial amount of diazonium salt present is proportional to V_∞. Hence $V_\infty - V_t$ is proportional to the amount of diazonium salt *present* in the reaction mixture at time t. The same proportionality constant applies throughout provided that the temperature and pressure of the collected nitrogen are constant.

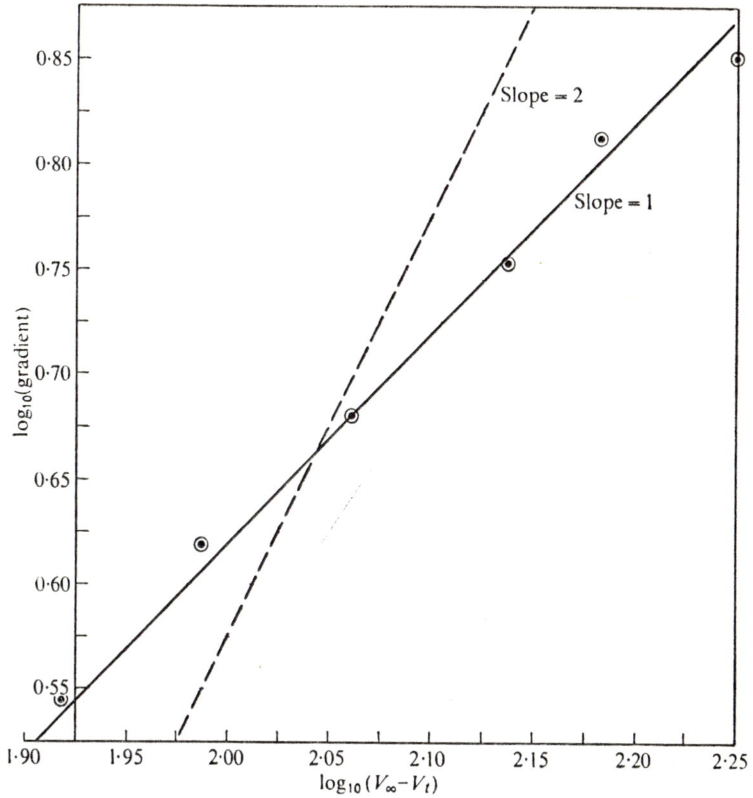

Fig. 4.4. A $\log_{10}$ (rate) against $\log_{10}$ (benzene diazonium chloride concentration) plot showing first order kinetics

Some of the results of an experiment are given in the first and second columns of Table 4.2. At six times during the reaction, the rate of reaction has been found by drawing tangents to the V_t against t curve .The gradients are given in column four of Table 4.2.

If the dependence of the rate of reaction on the concentration of benzene diazonium chloride has the form

$$-\frac{d[C_6H_5N_2Cl]}{dt} \propto [C_6H_5N_2Cl]^n$$

the value of n may be found by drawing a graph of $\log_{10}$ (rate of reaction) against $\log_{10}$ (concentration of benzene diazonium chloride) i.e. $\log_{10}$ (gradient of V_t versus t curve) against $\log_{10}(V_\infty - V_t)$. The graph is shown in Fig. 4.4, in which the full line has slope unity and the broken line has slope two.

If the order of the reaction is one, a graph of rate of reaction against concentration (gradient against $V_\infty - V_t$) should yield a straight line, whereas if the order is two, a graph of rate of reaction against concentration squared [gradient against $(V_\infty - V_t)^2$] should yield a straight line. These graphs are shown in Fig. 4.5.

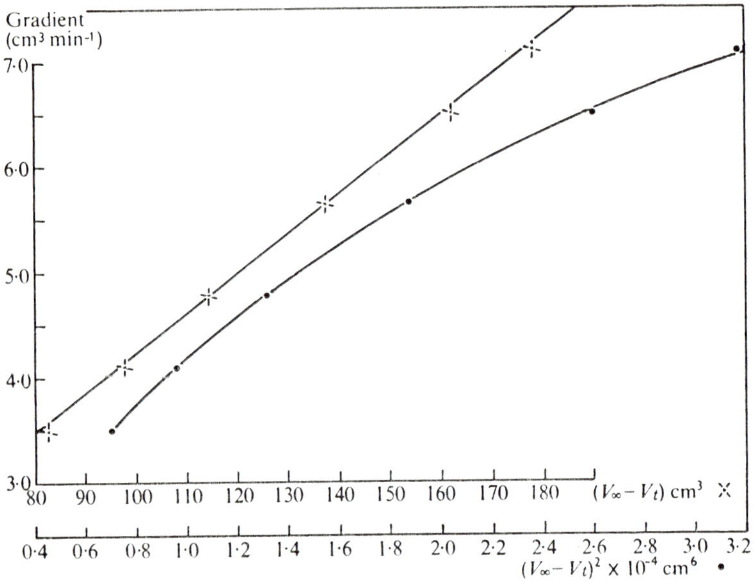

Fig. 4.5. Plots of rate of reaction against (concentration) and (concentration)² for the decomposition of benzene diazonium chloride

(b) Calculation of the order from integrated rate equations

If the reaction is first order with respect to benzene diazonium chloride

$$\log_e c = -k_1 t + \log_e c_0 \tag{3}$$

and

$$\log_{10} c = -\frac{k_1}{2{\cdot}30} . t + \log_{10} c_0$$

In terms of volume of nitrogen, this becomes

$$\log_{10}[q(V_\infty - V_t)] = -\frac{k_1 t}{2{\cdot}30} + \log_{10}[qV_\infty] \tag{21}$$

where q is the constant of proportionality which relates the volume of nitrogen to the amount† of benzene diazonium chloride which has decomposed. Equation (21) may be simplified to

$$\log_{10}(V_\infty - V_t) = -\frac{k_1}{2{\cdot}30}.t + \log_{10} V_\infty$$

by subtracting $\log_{10} q$ from both sides of the equation.

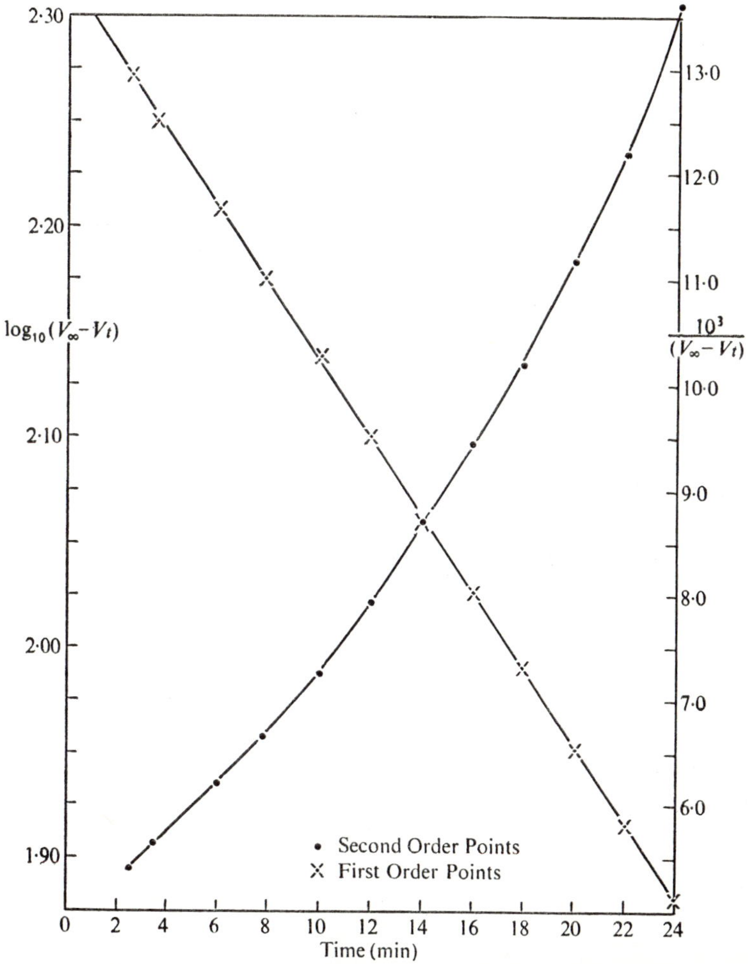

Fig. 4.6. First and second order plots for the decomposition of benzene diazonium chloride

† The concentration of benzene diazonium chloride is the amount in mol in the *fixed* volume of reaction mixture. See Appendix E.

If the reaction is second order with respect to benzene diazonium chloride then equation (10) applies

$$\frac{1}{c} = k_2 t + \frac{1}{c_0} \tag{10}$$

In terms of volumes of nitrogen, this becomes

$$\frac{1}{q(V_\infty - V_t)} = k_2 t + \frac{1}{qV_\infty}$$

or

$$\frac{1}{(V_\infty - V_t)} = qk_2 t + \frac{1}{V_\infty}$$

Provided that the reaction is either first or second order with respect to benzene diazonium chloride, a graph of *either* $\log_{10}(V_\infty - V_t)$ against time *or* $1/(V_\infty - V_t)$ against time will yield a straight line. The graphs are shown in Fig. 4.6.

It is fairly easy to detect first order kinetics without drawing a graph. Inspection of column three in Table 4.2 shows that the $\frac{1}{5}$th life (the time taken for the concentration of benzene diazonium chloride to decrease by one-fifth) is approximately 5 minutes, and does not depend on the value of $(V_\infty - V_t)$ chosen for its calculation. A constant fractional life throughout a reaction is characteristic of a first order reaction.

Alternatively, values for first and second order rate constants may be computed. Rearrangement of the integrated first order rate expression (3) gives

$$k_1 = \frac{2{\cdot}30}{t}\log_{10}\frac{c_0}{c} \tag{7}$$

From the integrated second order rate expression,

$$\frac{1}{c} - \frac{1}{c_0} = k_2 t \tag{10}$$

$$k_2 = \frac{c_0 - c}{c.c_0} \times \frac{1}{t}$$

Values for k_1 and k_2 are calculated at three times during the reaction in the table below.

Time min	$k_1 = \frac{2{\cdot}30}{t}\log_{10}\frac{c_0}{c} = \frac{2{\cdot}30}{t}\log_{10}\frac{V_\infty}{V_\infty - V_t}$ min^{-1}	$k_2 = \frac{c_0 - c}{c_0 c} \times \frac{1}{t} \propto \frac{V_t}{V_\infty(V_\infty - V_t)t}$ cm^{-3} min^{-1}
$3\frac{1}{2}$	$3{\cdot}0 \times 10^{-2}$	$1{\cdot}6 \times 10^{-4}$
12	$3{\cdot}7 \times 10^{-2}$	$2{\cdot}4 \times 10^{-4}$
24	$3{\cdot}9 \times 10^{-2}$	$3{\cdot}4 \times 10^{-4}$

The values for the first order rate constant, although varying by about 25 per cent, vary less than those for the second order rate constant†, indicating that the reaction is more likely to be first order than second.

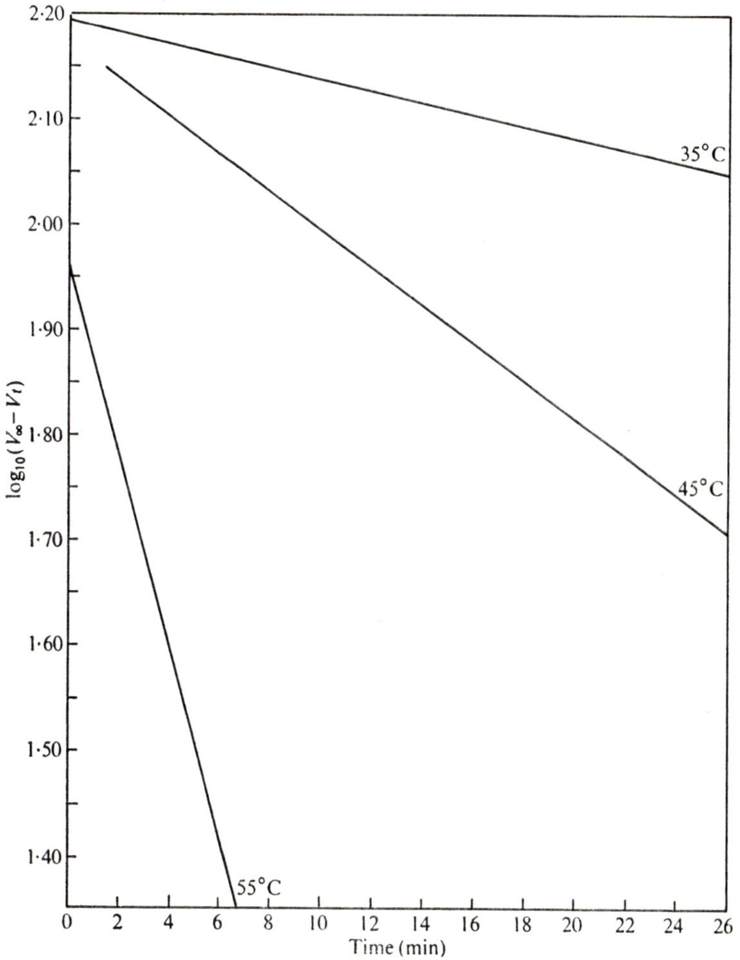

Fig. 4.7. First order plots for the decomposition of benzene diazonium chloride at three temperatures

When the order of the reaction has been found the rate constant is calculated most easily from the integrated rate expression graph. The first order graphs for the decomposition of benzene diazonium chloride at three temperatures are shown in Fig. 4.7. The gradients and rate constants are given below.

† The calculated constant is qk_2, not k_2

Temperature °C	*Gradient* min^{-1}	*Rate constant* k_1 min^{-1}
35	0·0056	0·013
45	0·018	0·041
55	0·090	0·21

Experiment 4.2 The iodination of acetone

When a solution of iodine in potassium iodide is acidified and mixed with acetone, monoiodoacetone is produced.

$$CH_3COCH_{3(aq)} + I_{2(aq)} = CH_3COCH_2I_{(aq)} + H^+_{(aq)} + I^-_{(aq)}\dagger$$

Contrast this reaction with the iodination of acetone in alkaline solution.

The kinetics of the reaction in acid solution are conveniently studied by measuring the concentration of iodine at known times during the reaction. A suitable reaction mixture is prepared (iodine, acetone, acid), samples are taken at known times and the reaction in the sample is quenched by neutralizing the acid with sodium hydrogen carbonate. (Iodination occurs fairly rapidly at both high and low pH, but only slowly when the pH is about seven.) The iodine remaining in the sample is then titrated with sodium thiosulphate solution. The rate of reaction is thus given by $-d[I_2]/dt$.

Unless very special concentration conditions are chosen, results of such experiments give information about the total order of the reaction, where the total order is defined as $(x + y)$ in the possible rate expression

$$-\frac{d[I_2]}{dt} = k_a[CH_3COCH_3]^x[I_2]^y$$

However, in the following experiments the conditions are chosen so that the order measured is not $(x + y)$. In Experiments (*a*) to (*d*), the reaction mixture contains a much larger concentration of acetone than iodine. Under these conditions, the change in the concentration of acetone during the reaction will be negligible compared with the change in the concentration of iodine. The suggested rate expression then becomes

$$-\frac{d[I_2]}{dt} = k_b[I_2]^y, \qquad \text{where } k_b = k_a[CH_3COCH_3]^x$$

Experiment (*a*) is designed to determine the value of y.

In Experiment (*b*), the concentration of acetone is still much higher than the concentration of iodine, but several runs are performed using different concentrations of acetone. In this way, the term $k_a[CH_3COCH_3]^x$ remains constant for any one run but varies from run to run depending on the value of the term $[CH_3COCH_3]^x$. Hence x can be determined.

Experiments (*c*) and (*d*) are designed to investigate two other interesting aspects of this reaction.

† In this reaction, the iodine is present mainly as the tri-iodide ion since the equilibrium constant for the reaction $I_{2(aq)} + I^-_{(aq)} \rightleftharpoons I^-_{3(aq)}$ is very large. The treatment of the results is unaffected mathematically when I_2 is substituted for I_3^-.

Reagents for all four experiments:

A. 0·05 M iodine in 0·4 M potassium iodide
B. 1·0 M hydrochloric acid
C. 0·10 M sodium hydrogen carbonate
D. 0·01 M sodium thiosulphate.

(*a*) *Determination of the order with respect to iodine*
Procedure

Into a conical flask (labelled 1) place
10 cm³ acetone (pipette)
10 cm³ solution B (pipette)
70 cm³ water (measuring cylinder).

Into each of several conical flasks (labelled 2, 3, etc.) pipette 10 cm³ of solution C.

To start the reaction, pipette 10 cm³ of solution A into flask 1, noting the time when the pipette is half empty. Shake well for 2 minutes. Withdraw 10 cm³ of the reaction mixture with a pipette and run this sample into flask 2, noting the time when the pipette is half empty. Mix well and titrate the residual iodine with solution D, adding starch near the end-point. At known times (5- to 7-minute intervals are suitable) repeat the titration of a 10 cm³ sample from flask 1, using flasks 3, 4, etc. Throw away the titration mixture and rinse the flasks well when the titration is complete. (Monoiodoacetone is lachrymatory.)

Tabulate the results and draw a graph of titre of residual iodine against time.

Discussion of results

Calculate the initial concentrations (mol dm^{-3}) of acetone and iodine in the reaction mixture. The density of acetone is about 0·8 g cm^{-3}. Show that when the iodine has completely reacted, the acetone concentration has changed by less than 0·5 per cent.

The graph obtained should be similar to that in Fig. 4.8. The concentration of iodine at a given time is proportional to the volume of sodium thiosulphate

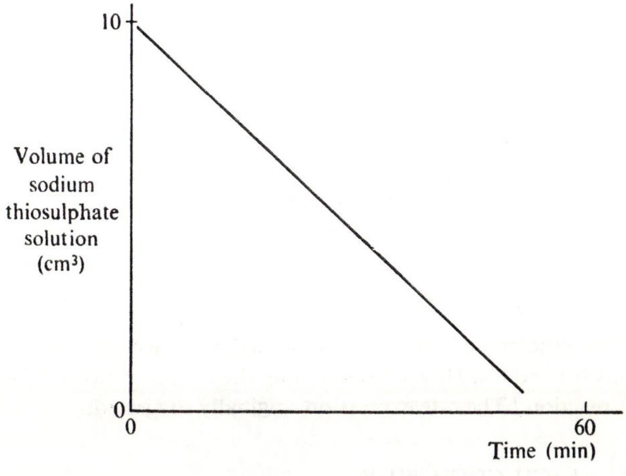

Fig. 4.8. Variation of [I_2] with time for Experiment 4.2 (*a*)

solution used in the titration at that time. The slope of the line in Fig. 4.8 is thus proportional to $-d[I_2]/dt$. Since the graph produces a straight line, $-d[I_2]/dt$ = constant. Thus the rate of reaction does not depend on the concentration of iodine and the value of y in the rate expression becomes zero. The reaction is said to be zero order with respect to iodine, i.e. $-d[I_2]/dt \propto [I_2]^0$.

(*b*) *Determination of the order with respect to acetone*

The procedure in this experiment is similar to that in Experiment (*a*) except that samples of the reaction mixture should be taken at about 3- to 4-minute intervals. The reaction mixture in flask 1 should contain:

20 cm^3 acetone (pipette);
10 cm^3 solution B (pipette);
60 cm^3 water (measuring cylinder).

Plot the results of Experiments (*a*) and (*b*) on the same graph and calculate the gradients of the lines. Experiment (*b*) may be repeated using other concentrations of acetone. (The total volume of reaction mixture should be 100 cm^3 in each case.)

Discussion of results

In Fig. 4.8 the gradient of the line is $k_a[CH_3COCH_3]^x$. For each run in Experiment (*b*) a straight line with this gradient will be obtained. If $x = 1$, doubling the concentration of acetone should double the gradient, whereas if $x = 2$, doubling the concentration of acetone should quadruple the gradient, and so on.

It can be shown that a graph of $\log_{10}$ (gradient) against $\log_{10}$ (volume of acetone in reaction mixture) should give a straight line with a gradient equal to x. Use this method to determine x, if its value is not apparent by inspection.

(*c*) *To determine the effect of changing the acid concentration*

Quantities. The mixture in flask 1 should contain:

10 cm^3 acetone (pipette);
20 cm^3 solution B (pipette);
60 cm^3 water (measuring cylinder).

Note that 20 cm^3 of solution C are needed to neutralize the acid.

Procedure

Proceed as in Experiment (*a*) taking samples at about 3- to 4-minute intervals. The experiment may be repeated using other concentrations of acid, but ensure always that the equivalent volume of solution C is used.

Discussion of results

Compare the gradient of the line(s) obtained with the gradients obtained in Experiments (*a*) and (*b*). How does doubling the concentration of the acid affect the rate of reaction? The rate expression originally suggested,

$$-\frac{d[I_2]}{dt} = k_a[CH_3COCH_3]^x[I_2]^y,$$

was modified to

$$-\frac{d[I_2]}{dt} = k_a[CH_3COCH_3]^x$$

after Experiment (*a*), and x has been determined in Experiment (*b*). What further modification is now necessary? Why is this last modification not important in Experiments (*a*) and (*b*)?

The results of Experiment (*c*) have been explained as catalysis of the reaction by hydrogen ions, but they are equally consistent with catalysis by chloride ions. How would you show that hydrogen ions, and not chloride ions, are responsible for the catalysis?

Hydrogen ions are formed in the reaction, and Experiment (*c*) shows that if the hydrogen ion concentration is increased, the rate of reaction rises. Thus the

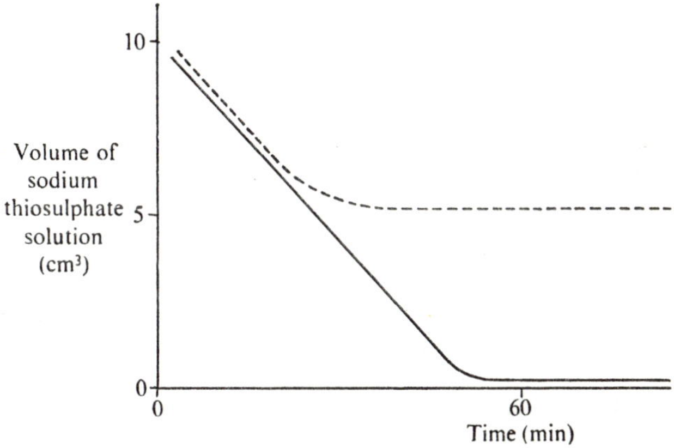

Fig. 4.9. Variation of $[I_2]$ with time for Experiments 4.2 (*a*) and (*d*)

straight lines in the graphs from Experiments (*a*), (*b*) and (*c*) should be curves. However, calculation shows that the curvature should be small. In Experiments (*a*) and (*b*), the hydrogen ion concentration increases by only 5 per cent during the whole reaction. Another factor is that, as the hydrogen ion concentration increases, the acetone concentration decreases. These two changes have opposite but unequal affects on the rate of reaction so that the 5 per cent error is reduced.

If one of the above experiments is allowed to proceed for long enough, the curve obtained is shown by the full line in Fig. 4.9. Why does the concentration of iodine remain constant after about an hour in Experiment (*a*), and after about half an hour in Experiments (*b*) and (*c*), and not decrease to zero? The following experiment should help to answer these questions.

> *(d) The effect of iodide ions*
>
> The procedure is the same as for (*a*), except that before the reaction is started 7·0 g of potassium iodide are dissolved in the contents of flask 1.

Discussion of results

The shape of the curve obtained for Experiment (*d*) is shown by the broken line in Fig. 4.9. Why does the concentration of iodine remain constant at a much higher level than in the previous experiments? From the results it is possible to calculate a value for the equilibrium constant of this reaction.

Clock reactions

Under the appropriate conditions the addition of an oxidizing agent to a solution containing iodide ions liberates iodine. The iodine reacts with excess iodide ions to form tri-iodide ions.

$$2I^-_{(aq)} - 2e = I_{2(aq)} \qquad \text{(oxidation of iodide ions)}$$

$$I_{2(aq)} + I^-_{(aq)} \rightleftharpoons I^-_{3(aq)}$$

Even when the concentration of iodine is low it can be detected easily by the addition of starch solution which produces an intense blue colour with iodine. If a small amount of sodium thiosulphate is added to a mixture of iodide solution and starch solution, then the iodine liberated by the later addition of the oxidizing agent reacts preferentially with the thiosulphate.

$$I_{2(aq)} + 2S_2O^{2-}_{3(aq)} = 2I^-_{(aq)} + S_4O^{2-}_{6(aq)}$$

The blue colour then appears only after all the thiosulphate has been used up.

Such a sequence of reactions forms the basis of a clock reaction. Three such reactions are investigated below.

Experiment 4.3 An iodine clock reaction

$$2I^-_{(aq)} + S_2O^{2-}_{8(aq)} = 2SO^{2-}_{4(aq)} + I_{2(aq)}$$

In this reaction the iodide ions are oxidized by persulphate ions.

Reagents: Solution A. 0·01 M potassium persulphate (0·68 g persulphate made up to 250 cm^3)
Solution B. 0·3 M potassium iodide + 0·0005 M sodium thiosulphate + starch (12·5 g potassium iodide, 0·03 g sodium thiosulphate, 10 cm^3 starch, made up to 250 cm^3)

Procedure

Leave both solutions standing in the laboratory for some time to achieve room temperature. Check with a thermometer.

Using a measuring cylinder pour 25 cm^3 samples of solutions A and B into separate 100 cm^3 beakers placed on a piece of white paper. Pour solution A into solution B and, at the same time, start the stop-clock. To ensure thorough mixing, pour rapidly from one beaker to the other, twice. Record the time when the blue colour appears.

Repeat the experiment, varying the concentration of the persulphate solution. Use 25 cm^3 of solution B in each case.

Beaker 1	*Beaker* 2		
Solution B cm^3	*Solution* A cm^3	*Water* cm^3	*Time t* s†
25	25	0	43
	20	5	55
	15	10	71
	10	15	100

Discussion of results

Is there any obvious relationship between the times, t, and the volumes of persulphate solution used? If there is, try to plot a graph which will give a straight line.

What determines the time taken for the blue colour to appear? If the rate of oxidation of iodide ions is high, the thiosulphate is soon consumed and the blue colour appears in a short time. If the rate of oxidation of iodide ions is low, the blue colour appears only after relatively longer time. The average rate of oxidation of iodide ions during this time interval is proportional to the reciprocal of the time. If only a very small fraction of the total quantity of iodine which would be produced on completion of the reaction is required to give the blue colour, then this *average* rate will be a good estimate of the *initial* rate of the reaction. This condition can be achieved by using a very small quantity of thiosulphate compared with iodide and persulphate.

The advantage of using initial rates is that they depend only on the initial concentrations of the reactants, which we know and can easily vary (as in Experiment 4.3). This situation is analogous to that in the thiosulphate–acid reaction of Experiment 3.2.

From the molarities and quantities given in Experiment 4.3 it is possible to calculate the quantity of iodine produced before the blue colour appears, since this is controlled by the amount of thiosulphate ions present. Verify that this amount of iodine is about 2·5 per cent of the total amount of iodine which would be produced by the persulphate–iodide reaction if it went to completion. This percentage is low enough for $1/t$ to be considered proportional to the initial rate. Values of $1/t$ obtained in the experiment should be related in some way to the persulphate concentration. Since the iodide concentration and the amount of thiosulphate are kept constant, the initial rate will not be affected by these. It should now be clear how to interpret any graphs plotted from the experimental results.

Theory of clock reactions

Most clock reactions involve three steps.

1. *The main reaction*

e.g. $A + B = C + D$

(Equivalent to the oxidation of iodide ions by persulphate ions in Experiment 4.3.)

† These are results of an experiment performed at 20°C.

This is the reaction whose kinetics are being studied, and it must be slow compared with reactions 2 and 3.

2. *The monitor reaction*

e.g. $C + M = \text{products}$

(In Experiment 4.3, C is iodide ions and M is thiosulphate ions.) This is the reaction between one of the products of the main reaction and a *monitor* substance, M. The initial concentration of M must be considerably less than those of the main reactants, A and B. When all the monitor has been used up, a third, or indicator, reaction occurs. Like the monitor reaction this is relatively fast.

3. *The indicator reaction*

e.g. $C + In' = In''$

where In′ and In″ are the two indicator colours involved. In Experiment 4.3 In′ is colourless and In″ blue. The starch–iodine reaction is commonly used in clock reactions because the blue colour appears rapidly and is easy to observe. It has the added advantage of being reversible so that, on addition of more monitor substance, the blue colour disappears and the 'clock' can be 'rewound'. (See Experiment 4.5.)

Two further clock reactions are described below, each of which is used to obtain a possible rate expression for the main reaction.

Experiment 4.4 A bromine clock reaction

Main reaction. $6H^+_{(aq)} + 5Br^-_{(aq)} + BrO^-_{3(aq)} = 3Br_{2(aq)} + 3H_2O_{(l)}$
Monitor reaction. $3Br_{2(aq)} + C_6H_5OH_{(aq)} = HO.C_6H_2.Br_{3(aq)} + 3H^+_{(aq)} + 3Br^-_{(aq)}$
Indicator reaction. $Br_{2(aq)}$ + methyl orange = colourless compound.

Here the 'indicator' behaves as a destructible dye.

The effect of varying the concentration of each reactant is investigated by keeping two concentrations constant whilst altering the third. The concentration of monitor substance is chosen so that only 1–2 per cent of the complete main reaction occurs before the methyl orange is decolourized.

(*a*) *Variation of initial rate with bromide and bromate ion concentrations*

Reagents: 0·005 M potassium bromate;
0·01 M potassium bromide;
0·001 M phenol.
Solution C. 40 cm³ 1 M sulphuric acid and 2·0 cm³ bench methyl orange solution (5g solid in 1 dm³ of water) made up to 100 cm³ with pure water (the solution is about 0·4 M with respect to the acid).

Just before beginning the experiment, dilute 5·0 cm³ of the phenol solution to 100 cm³ using pure water, and use this diluted solution as directed in the table below. (The phenol is now about 0·00005 M.)

Quantities: Varying the bromide ion concentration

Beaker 1		*Beaker 2*				
Vol. of KBr cm^3	*Vol. of* H_2O cm^3	*Vol. of* $KBrO_3$ cm^3	*Vol. of* *Soln.* C cm^3	*Vol. of* *phenol* cm^3	t† s	$\frac{1}{t}$ s^{-1}
10·0	0	10·0	15·0	1·0	37·0	
8·0	2·0	10·0	15·0	1·0	45·0	
6·0	4·0	10·0	15·0	1·0	56·0	
5·0	5·0	10·0	15·0	1·0	65·0	
4·0	6·0	10·0	15·0	1·0	82·0	
3·0	7·0	10·0	15·0	1·0	108·0	

Varying the bromate ion concentration

Use the same volumes of phenol and solution C as above, but keep the volume of bromide solution constant at 10·0 cm^3 and vary the volume of bromate solution as follows.

Beaker 1			
Vol. of $KBrO_3$ cm^3	*Vol. of* H_2O cm^3	t† s	$\frac{1}{t}$ s^{-1}
10·0	0	37·0	
8·0	2·0	44·5	
6·0	4·0	57·0	
5·0	5·0	71·0	
4·0	6·0	89·0	
3·0	7·0	119·0	

Procedure

Allow the solutions to stand for some time so that they reach room temperature. Check this using a thermometer. Use separate burettes to measure the phenol solution, water, and the solution whose volume is being varied. The other solutions can be dispensed from suitable pipettes or measuring cylinders. Pour the solutions in beaker 1 into the solutions in beaker 2 and at the same time start the stop-clock. To ensure thorough mixing, pour rapidly from one beaker to the other twice. Record the time when the pink colour of the methyl orange has *completely* disappeared and the solution has become colourless. Some practice will be needed to standardize this procedure.

(*b*) *Variation of initial rate with the hydrogen ion concentration*

Reagents: 0·01 M sulphuric acid.
0·2 M potassium bromate.
0·00005 M phenol (from the above experiments).

† These are the results of experiments performed at 15°C.

Solution D. 250 cm^3 containing 12·0 g potassium bromide and 5·0 cm^3 bench methyl orange solution made up with pure water. (The solution is about 0·4 M with respect to potassium bromide.)

Beaker 1			*Beaker 2*			
Vol. of H_2SO_4 cm^3	*Vol. of* H_2O cm^3	*Vol. of* $KBrO_3$ cm^3	*Vol. of Soln.* D cm^3	*Vol. of phenol* cm^3	t† s	$\frac{1}{t}$ s^{-1}
10·0	0	10·0	15·0	1·0	11·5	
8·0	2·0	10·0	15·0	1·0	16·5	
6·0	4·0	10·0	15·0	1·0	29·5	
5·0	5·0	10·0	15·0	1·0	45·0	
4·0	6·0	10·0	15·0	1·0	69·5	
3·0	7·0	10·0	15·0	1·0	125·0	

Follow the procedure given for the variation in the bromide ion concentration.

Discussion of results

If the rate expression can be assumed to be of the form

$$+\frac{d[Br_2]}{dt} \propto [H^+]^x[BrO_3^-]^y[Br^-]^z$$

where x, y and z are the orders with respect to H^+, BrO_3^- and Br^- respectively, then

$$\text{Initial rate} \propto [H^+]_0^x[BrO_3^-]_0^y[Br^-]_0^z$$

where these are the initial concentrations. For the variation in $[H^+]_0$, the $[BrO_3^-]_0$ and $[Br^-]_0$ were kept constant so that

$$\text{Initial rate for } [H^+]_0 \text{ variation} \propto [H^+]_0^x$$

Taking logarithms,

$$\log_{10}(\text{initial rate}) = x\log_{10}[H^+]_0 + \log_{10}(\text{a constant})$$

A plot of $\log_{10}$ (initial rate) against $\log_{10}[H^+]_0$ should be a straight line of slope x. Similar arguments apply to the $[BrO_3^-]_0$ and $[Br^-]_0$ variation.

The experimental data provide two terms; one term is proportional to the initial rate, and one is proportional to the concentration of the reactant being varied. Plot the required graph and determine the orders x, y and z. Are these values integers? If they seem to be, further plots can be made of initial rates against concentrations raised to the appropriate power. Such plots should give straight lines through the origin for integral orders. If any other lines (or curves) are obtained there are many possible interpretations, one being that the order is fractional.

What is the *total* order of the reaction?

† These are results from an experiment performed at 20°C.

Experiment 4.5 The Harcourt-Esson reaction

Main reaction. $H_2O_{2(aq)} + 2H^+_{(aq)} + 2I^-_{(aq)} = I_{2(aq)} + 2H_2O_{(l)}$†.
Monitor reaction. $I_{2(aq)} + 2S_2O^{2-}_{3(aq)} = 2I^-_{(aq)} + S_4O^{2-}_{6(aq)}$.
Indicator reaction. $I_{2(aq)}$ + starch $\rightleftharpoons$ blue substance.

Since the indicator reaction is reversible this 'clock' can be 'rewound' by the addition of more sodium thiosulphate. The blue starch–iodine colour is discharged and the main reaction, which continues uninterrupted, eventually produces sufficient iodine to use up the second quantity of monitor so that the blue colour reappears. Successive additions of monitor can be made until the main reaction has been completed and no more iodine is being produced.

In this experiment, the effects of varying the peroxide concentration and the iodide ion concentration are investigated. The rate expression obtained will be incomplete since it will not include any information concerning the effect of varying the hydrogen ion concentration, which is a difficult investigation to carry out. The concentrations of all three reactants can vary during a 'run', but the conditions chosen are such that two remain effectively constant. The rate of reaction during any one run then depends only on the third.

(*a*) *Order with respect to hydrogen peroxide*

Reagents: 0·06 M potassium iodide (make up 1 dm³ and keep the excess for the variation in iodide concentration)
sulphuric acid (1 part concentrated acid, 1 part water by volume)
'1 vol' hydrogen peroxide
0·1 M sodium thiosulphate
starch solution

Quantities (for one run):

200 cm³ potassium iodide (in 250 cm³ conical flask)
10·0 cm³ sulphuric acid }
10·0 cm³ starch } in 100 cm³ flask
10·0 cm³ hydrogen peroxide }

Procedure

Measure the three smaller volumes accurately using pipettes or burettes, and use a measuring cylinder for the iodide solution.

It is important that the temperature of the reactants remains constant throughout the experiment. Keep the flasks immersed in a trough of cold water for a few minutes before beginning the experiment. Regularly immerse the reaction vessel in the same trough during the 'run'.

Add 1·0 cm³ of thiosulphate solution to the iodide solution from a burette. Mix the reactants and, at the same time, start the stop-clock. Shake vigorously to ensure thorough mixing. Record the time when the blue colour appears. Immediately add another 1·0 cm³ of thiosulphate solution and shake again. Note the time when the blue colour reappears. Repeat the addition of 1·0 cm³ portions of the monitor until a total volume of 14·0 cm³ has been added.

† The iodine is present mainly as $I^-_{3(aq)}$ ions.

Measurement of V_∞

The volume of the monitor (V_∞) required to react with the total amount of iodine produced by the main reaction could be found by allowing the reaction to go to completion. A quicker method is to use the same initial quantity (i.e. mol) of hydrogen peroxide but to concentrate the other solutions so that the reaction is over very quickly. This can be done because only the hydrogen peroxide is completely consumed during the reaction so that the yield of iodine depends only on the initial amount of this reactant.

Add about 10 cm³ of concentrated sulphuric acid to 10·0 cm³ of '1 vol' hydrogen peroxide from (*a*). (CARE) Dissolve about 8 g of potassium iodide in the resulting solution. If necessary, add the minimum quantity of water to complete the solution of the solid. Shake, and titrate with 0·1 M sodium thiosulphate to find V_∞, using starch as indicator near the end point.

What is the theoretical value of V_∞ given by the initial quantities and concentrations? How close is this to the experimental value?

Discussion of results

Some experimental results are given below.

Vol. of thiosulphate† V_t cm³	*t* s	($V_\infty - V_t$) cm³
1·0	7·0	16·0
2·0	18·0	15·0
3·0	33·0	14·0
4·0	48·5	13·0
5·0	66·0	12·0
6·0	85·0	11·0
7·0	107	10·0
8·0	131	9·0
9·0	157	8·0
10·0	188	7·0
11·0	222	6·0
12·0	264	5·0
13·0	313	4·0
14·0	372	3·0

$V_\infty = 17{\cdot}0$ cm³.
Temperature = 23°C.
Quantities and concentrations as above.

Why do the time intervals between the reappearance of the blue colour lengthen as the reaction proceeds?

At a time *t*,

$$[H_2O_2]_{used} = [I_2]_{produced} \propto V_t$$

† Each addition of thiosulphate corresponds to an equal amount of iodine.

also

$$[H_2O_2]_{\text{initial}} \propto V_\infty$$

Therefore

$$[H_2O_2]_{\text{remaining at time } t} \propto (V_\infty - V_t)$$

It may be observed that this latter relationship is like the one derived in Experiment 3.4 relating peroxide concentration to $(V_\infty - V_t)$ where V represents a *gas* volume. The essential theory of the two experiments is similar and the titration readings in 4.5 are equivalent to the gas volumes in 3.4.

There are two ways of analysing the experimental data to determine the order with respect to hydrogen peroxide.

1. Plot a graph of V_t against t and draw tangents to the curve at suitable points. Find the slope of each tangent and plot $\log_{10}$(tangent slope) against $\log_{10}(V_\infty - V_t)$, as in Experiment 3.4. The slope of the latter plot will be the order with respect to hydrogen peroxide.
2. Plot first and second order graphs as given in Table 4.1. The simplest plots are $\log_{10}[H_2O_2]$ against t and $1/[H_2O_2]$ against t respectively, where $[H_2O_2]$ can be represented by $(V_\infty - V_t)$. A straight line for one of these plots will indicate the order.

If the rate expression can be assumed to be of the form

$$+\frac{d[I_2]}{dt} \propto [H_2O_2]^x[I^-]^y[H^+]^z$$

then, when $[I^-]$ and $[H^+]$ are kept constant,

$$+\frac{d[I_2]}{dt} \propto [H_2O_2]^x$$

and

$$+\frac{d[I_2]}{dt} = k'[H_2O_2]^x$$

Note that k' is a special form of the rate constant since it contains terms in $[H^+]$ and $[I^-]$. Evaluate k' for the graphs in 2, being careful to quote its units. For the $\log_{10}(V_\infty - V_t)$ against t plot, for instance, the slope (if it were a straight line) would be equal to $-k'/2{\cdot}3$, and the units of k' would be s^{-1} because no concentration units appear in a first order rate constant.

Errors

Most sources of error in this experiment have been minimized. The temperature is kept constant, measuring errors should be small, and the appearance of the blue colour is sudden. However, the progressive dilution of the reaction mixture by the thiosulphate solution has to be taken into account. What effect will this have on the reaction rate? In view of this, why was such a large volume of iodide solution used? Calculate the maximum percentage error likely to be caused by this effect. Is it large enough to affect the graphs seriously? If it is, could it be allowed for on the graphs?

Do the hydrogen ion concentration and the iodide ion concentration remain effectively constant? In the main reaction, two iodide ions are consumed for every one hydrogen peroxide molecule, but in the monitor reaction the iodine molecule produced in the main reaction is converted to two iodide ions. The concentration of iodide ions therefore does remain constant throughout the reaction if the monitor reaction is very fast compared with the main reaction.

After the reaction mixture has been prepared, the concentrations of hydrogen ions and hydrogen peroxide are about 0·4 M and 0·004 M respectively, whilst the iodide concentration is about 0·05 M. There is therefore an excess of both hydrogen and iodide ions, and in this situation the hydrogen peroxide is said to be *isolated.* All the peroxide reacts when the reaction goes to completion so that its concentration changes by 100 per cent. During this time, however, the percentage change in the hydrogen ion concentration is very small and it is assumed that the effect of this change on the reaction rate is negligible compared with the effect of the changing peroxide concentration. The percentage change in the hydrogen ion concentration can be calculated using the above figures. Show that this is only 2 per cent during the reaction.

(b) Order with respect to iodide ions

In any one run the iodide concentration remains constant so that the results in (*a*) indicated nothing about the order with respect to iodide ions.

If it is possible to write a rate expression of the form

$$+\frac{d[I_2]}{dt} \propto [H_2O_2]^x[I^-]^y[H^+]^z,$$

then it has been shown already that, for one run,

$$+\frac{d[I_2]}{dt} \propto [H_2O_2]^x$$

and

$$+\frac{d[I_2]}{dt} = k'[H_2O_2]^x$$

where k' is a constant and equal to $k''[I^-]_0^y[H^+]_0^z$. ($[I^-]_0$ and $[H^+]_0$ are the initial concentrations of iodide ions and hydrogen ions and remain effectively constant during a run as shown above.)

If the initial concentration of iodide ions is varied from run to run whilst $[H^+]_0$ is kept constant along with the peroxide concentration, then

$$k' \propto [I^-]_0^y$$

If $y = 1$, then $k' \propto [I^-]_0$, and if $y = 2$, then $k' \propto [I^-]_0^2$. It should be emphasized that during the run it is still only the peroxide concentration which varies so that the order of reaction remains the same as that determined in (*a*). However, the value of k' for each run is affected by $[I^-]_0$.

Reagents: These are the same as in (*a*).
Quantities: Vary the iodide concentration as follows:

Vol. KI *soln.* cm^3	*Vol.* H_2O cm^3
175	25
150	50
125	75
100	100

[Run (*a*) will provide readings for 200 cm^3 potassium iodide solution and 0 cm^3 water]

Procedure

This is as run (*a*), but it is only necessary to make about six 1·0 cm^3 additions of thiosulphate solution. Because the same volume and concentration of peroxide is used as in (*a*), and because the acid and iodide are always in excess, V_∞ is the same as in (*a*).

Discussion of results

Some results are plotted in Fig. 4.10. Each slope is equal to $-k/2{\cdot}3$ for a first order plot so that k' can be calculated.

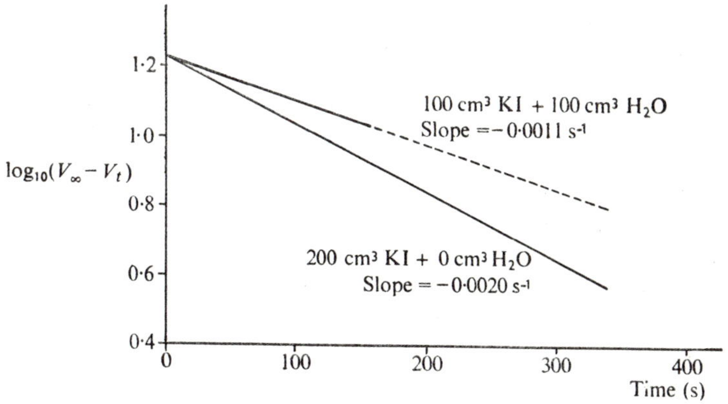

Fig. 4.10. Plots of $\log_{10}(V_\infty - V_t)$ against t for two different initial concentrations of iodide at 23°C. $V_\infty = 17{\cdot}0\ cm^3$

Plot any experimental results obtained in (*b*) on a graph similar to Fig. 4.10. Determine the slopes of each of the five straight lines and construct a table of iodide volumes used and calculated values of k'. Since $k' \propto [I^-]_0^y$, what is the value of y to the nearest whole number? A plot of $\log_{10} k'$ against $\log_{10}[I^-]_0$ should be drawn to determine an accurate value for y which is given by the slope of the straight line. Decide whether or not y can be considered an integer within experimental error.

Practical exercises

1. Investigate the kinetics of the clock reaction between formaldehyde and hydrogen sulphite ions using phenolphthalein as the indicator. (See *S.S.R.*, 1956, **38**, 100.)
2. Investigate the rate of hydration of carbon dioxide in the reaction known as the 'soda water clock'. (See *S.S.R.*, 1964, **157**, 532.)
3. Investigate the rate of decomposition of hydrogen peroxide when catalysed by colloidal manganese(IV) oxide.

Add 1·0 cm³ of 0·02 M potassium permanganate solution from a burette to 25·0 cm³ of '5 vol' hydrogen peroxide, and start the stop-clock. After 15 seconds pour the reaction mixture into 224 cm³ of pure water to quench the reaction. Withdraw 25 cm³ samples and titrate against 0·02 M permanganate. Repeat the whole procedure several times, quenching the reaction at 30, 45, 60 seconds, and so on up to about 105 seconds, obtaining titres V_t for each time. Obtain the titre at $t = 0$ by adding the 1·0 cm³ of permanganate and 25·0 cm³ of hydrogen peroxide to 224 cm³ of pure water without first mixing them.

Plot a graph of $\log_{10} V_t$ against time, t. What is the order of reaction with respect to hydrogen peroxide?

4. Investigate the kinetics of the reaction between iodide ions and persulphate ions using a volumetric method.

$$2I^-_{(aq)} + S_2O^{2-}_{8(aq)} = I_{2(aq)} + 2SO^{2-}_{4(aq)}$$

(*a*) To determine the total order of the reaction, the initial reaction mixture should contain 50·0 cm³ of 0·120 M potassium iodide solution and 50·0 cm³ of 0·060 M potassium persulphate solution. Titrate the 10 cm³ samples with 0·020 M sodium thiosulphate solution. Do not continue sampling after iodine has begun to precipitate. *Calculate* the value for T_∞, the titre after the reaction has been completed. (This cannot be determined experimentally. Why not?)

(*b*) Determine the order with respect to persulphate ions using an initial reaction mixture of 100 cm³ of 0·60 M potassium iodide solution and 100 cm³ of 0·015 M potassium persulphate solution. Titrate 10 cm³ samples against 0·01 M sodium thiosulphate solution. T_∞ can be determined experimentally in this case. (Why?)

Plot suitable graphs to determine the order of reaction in (*a*) and in (*b*). Comment on the values obtained.

5 Other experimental techniques for determining reaction rates

A. The use of variations in physical properties of the reaction mixture to follow concentration changes

Very few methods of investigating the kinetics of a reaction involve direct measurement of concentration. Usually some property related to concentration, which can readily be measured, is used. Some properties vary *linearly* with the concentration changes of the reaction mixture. A few, for example colour intensity, do not have a linear dependence on concentration, and in these cases it is most convenient to plot a calibration curve so that readings of the property can be directly converted into concentrations. For a linear relationship, it is not necessary to plot a calibration curve as the following treatment shows.

Consider the reaction whose stoichiometric equation is:

$$A = B$$

Let the property being measured be P_t at time, t. The rate of reaction is given by:

$$-\frac{d[A]_t}{dt} = -\frac{dP_t}{dt} \times \frac{d[A]_t}{dP_t} \tag{1}$$

If P varies linearly with [A], then:

$$[A]_t = b + aP_t \tag{2}$$

where a and b are constants under the experimental conditions. Therefore,

$$\frac{d[A]_t}{dP_t} = a$$

and from equation (1)

$$-\frac{d[A]_t}{dt} = -\frac{dP_t}{dt} \times a \tag{3}$$

A graph of P_t (or a function of P_t) will have the same form as a graph of $[A]_t$ (or the same function of $[A]_t$) against t. This indicates that the order of reaction can be deduced from graphs involving P_t as well as from graphs involving $[A]_t$; there is no need to convert dP_t/dt to $d[A]_t/dt$ first.

However, if it is required to obtain a value for a rate constant it is sometimes necessary to make such a conversion. For a first order reaction the units of k_1 are time^{-1} so that it cannot matter whether values of P_t or $[A]_t$ are plotted. But in the case of a reaction whose order is 2 or more the units of the rate constant are $(\text{mol}^{-1}\ \text{dm}^3)^n\ \text{s}^{-1}$, where n is a small integer. It is therefore necessary to make the conversion from dP_t/dt to $d[A]_t/dt$. To do this, the value of the constant in equation (3) must be determined. This can be shown to be $([A]_\infty - [A]_0)/(P_\infty - P_0)$, where subscripts 0 and ∞ indicate initial and final values respectively. For an irreversible reaction $[A]_\infty = 0$ so that $a = -[A]_0/(P_\infty - P_0)$. Equation (1) can thus be written as:

$$-\frac{d[A]_t}{dt} = \frac{dP_t}{dt} \times \frac{[A]_0}{P_\infty - P_0} \tag{4}$$

Some properties which can be used

The volume (V_t) of a gas evolved in a reaction at constant pressure has already been used to follow the course of a reaction (Experiments 3.4 and 4.1). The order of reaction was determined by plotting V_t against time and drawing tangents to the curve, or by plotting either $\log_{10}(V_\infty - V_t)$ or $1/(V_\infty - V_t)$ against time. There was no need to convert values of V_t to the corresponding reactant concentrations.

Volumetric methods such as Experiment 4.2 do not directly measure concentrations since the measurements recorded are titres, i.e. volumes of a standard solution. Again, however, the titres were used for plotting suitable graphs so that the order could be determined.

Many physical properties which are related in some way to the concentration in the reacting mixtures can be used to follow the course of a reaction. A list is given below of the most important of these. In many cases a linear relationship does exist (as in Experiments 5.1 and 5.2), but this is not so for them all (see Experiment 5.3).

1. Electrical conductance.
2. Refractive index or viscosity of liquids.
3. Thermal conductivity of gases.
4. Pressure of a gas at constant volume.
5. Colour intensity.
6. Density of a liquid.
7. Specific rotation of the plane of polarized light by optically active solutions.

The use of properties 5, 6, and 7 is illustrated in Experiments 5.3, 5.2, and 5.1 respectively.

Experiment 5.1 The inversion of sucrose

Sucrose is hydrolyzed in aqueous solution according to the equation:

$$\underset{\text{sucrose}(+)}{C_{12}H_{22}O_{11(aq)}} + H_2O_{(l)} = \underset{\text{glucose}(+)}{C_6H_{12}O_{6(aq)}} + \underset{\text{fructose}(-)}{C_6H_{12}O_{6(aq)}}$$

When plane polarized light from a monochromatic source is passed through sucrose solution, the plane of polarization is rotated through a certain angle. Sucrose and glucose are dextrorotatory, whilst fructose rotates the light in the opposite direction and is thus laevorotatory. However, fructose is *strongly* laevorotatory and glucose is *weakly* dextrorotatory. As the concentration of the products increases, the large positive (i.e. dextro-) rotation of sucrose is changed to the net laevo-rotation of the product mixture.

The reaction is carried out in acid solution because hydrogen ions act as a catalyst for the hydrolysis (see Table 8.3).

Quantities: weight of sugar 10·00 g
volume of pure water 16·0 cm^3
volume of concentrated hydrochloric acid 4·0 cm^3

Procedure

For detailed instructions on the use of a polarimeter, the reader should consult a standard textbook of practical physical chemistry. Care should be taken to ensure that the reaction mixture has achieved thermal equilibrium before beginning each run. In run 1, readings of the angle of rotation should be recorded every few minutes for at least 1 hour. The angle of rotation at the end of the reaction (R_∞) is also required.

Discussion of results

Some results using the above quantities and concentrations are given in Table 5.1.

Table 5.1

R_t degree	+43·5	+35·0	+27·0	+14·0	+3·8	−2·6	−6·7
t min	0	10	20	40	60	80	100

R_∞ (after 12 hours) = −16·0 degree.
Temperature = 19°C.

These results can be analysed in the usual ways. [Note that the concentration of sucrose at any given time is proportional to ($R_t - R_\infty$).] The order of the reaction for each run should be *one*. What is the significance of this observed order? There are two possible rate expressions which could be written and this situation

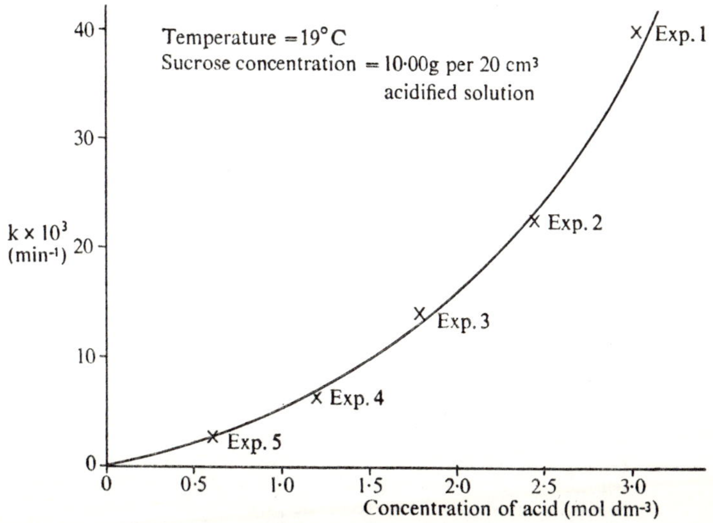

Fig. 5.1. Variation of the first order rate constant with the acid concentration for the hydrolysis of sucrose

is similar to that in the acid-catalysed hydrolysis of methyl acetate (Experiment 9.6), and in Experiment 5.2.

The dependence of the rate on the concentration of catalyst can be investigated by varying the volume of acid whilst keeping the total volume of the reaction mixture constant. Fig. 5.1 is a plot of the first order rate constant in each experiment (calculated from the slope in the first order graph) against the molarity of the acid. What relationship between the rate and the concentration of the catalyst is suggested by this plot? Compare this relationship with that obtained in Experiments 9.6 and 5.2.

Experiment 5.2 The depolymerization of diacetone alcohol

$$(CH_3)_2C(OH)CH_2COCH_{3(aq)} = (CH_3)_2CO_{(aq)} + CH_3COCH_{3(aq)}$$

In aqueous solutions, the depolymerization of diacetone alcohol to form acetone is reversible, but the concentration of diacetone alcohol at equilibrium is so small that the reaction may be assumed to proceed to completion. It is found that, whereas the reaction is catalysed by hydroxide ions, it is not catalysed by other (Brönsted and Lowry) bases and acids. It is therefore an example of a reaction which exhibits specific catalysis by hydroxide ions (see Table 8.3).

The reaction is accompanied by an increase of volume and the rate of reaction may be determined by measuring the change of volume of the reaction mixture during the reaction. A dilatometer is used to measure the small (about 5 per cent) volume changes.

Design and use of a dilatometer

A dilatometer is a reaction vessel which is fitted with a tap through which it can be filled and a capillary tube through which the volume increase can occur and be measured. All-glass dilatometers can be obtained but it is possible to make one with easily obtainable apparatus. For example, a separating funnel may be fitted with a bung and capillary tube.

The dilatometer is firmly clamped and is then filled through the tap by sucking through a rubber tube which is attached to the capillary tube. When the liquid has just entered the capillary tube, the tap is closed. The position of the meniscus in the capillary tube can be measured with a suitably clamped metre rule.

It is important that no air bubbles remain in the dilatometer and the capillary tube should be rinsed and dried before each experiment by sucking water, alcohol, ether and then air through it with a tap pump.

Procedure

The experimental directions and results given below are for a dilatometer with a capacity of about 80 cm^3 and a capillary tube with a bore of 2·5 mm.

To 100 cm^3 0·100 M sodium hydroxide solution add 5·0 cm^3 diacetone alcohol, mix them well and fill the dilatometer. Read the position of the meniscus (l_t) in the capillary tube every two minutes for about forty minutes. Allow the apparatus to stand until the reaction has stopped and read the position of the meniscus (l_∞).

A dilatometer is also a thermometer: precautions should therefore be taken to maintain a constant temperature during the run, and when the l_∞ reading is taken.

Discussion of results

Some results are given in Table 5.2.

Table 5.2. The depolymerization of diacetone alcohol in 0·10 M sodium hydroxide solution at 17°C

time min	4	8	12	16	24	32	40
l_t cm	7·4	13·2	18·3	23·0	31·1	37·7	43·7

$l_\infty = 63{\cdot}5$ cm

Determine the order of the reaction by plotting a suitable function of $(l_\infty - l_t)$ against t and determine the rate constant for the reaction. Decide to which reactant this order must refer.

To determine the order with respect to the catalyst, Experiment 5.2 may be repeated using different concentrations of sodium hydroxide solution. It is found that the first order rate constant is directly proportional to the hydroxide ion concentration. The total order of the reaction is thus *two* and the rate expression is:

$$-\frac{d[\mathrm{D}]}{dt} = k_2[\mathrm{OH^-}][\mathrm{D}]$$

where D represents diacetone alcohol. The constant, k_2, is the second order rate constant, and the first order rate constant calculated in any one run is thus equal to $k_2[\mathrm{OH^-}]$.

The determination of the dissociation constants of weak acids and weak alkalis

In Experiment 5.2 it was found that

$$k_1 = k_2[\mathrm{OH^-}] \qquad (5)$$

where k_2 is the second order rate constant. Similar results are obtained if other strong alkalis are used and the value of k_2 is found to be constant at constant temperature.

This result is important because it means that the reaction may be used to determine the concentration of hydroxide ions in a solution, and hence the dissociation constants of weak alkalis.

Example

Experiments with potassium hydroxide and thallium(I) hydroxide gave the following results.

The first order rate constant using 0·219 M potassium hydroxide was found to be 0·0477 $\mathrm{min^{-1}}$ (decadic logs), whereas in 0·219 M thallium(I) hydroxide solution, the value is 0·0370 $\mathrm{min^{-1}}$. Potassium hydroxide is believed to be completely dissociated so that

$$k_2 = \frac{k_1}{[\mathrm{OH^-}]} = \frac{0{\cdot}0477}{0{\cdot}219} = 0{\cdot}218\ \mathrm{dm^3\ mol^{-1}\ min^{-1}}$$

Using equation (5) to calculate the concentration of hydroxide ions in the thallium(I) hydroxide solution

$$[OH^-] = \frac{0{\cdot}0370}{0{\cdot}218} = 0{\cdot}170 \text{ mol dm}^{-3}$$

Now $TlOH_{(aq)} \leftrightharpoons Tl^+_{(aq)} + OH^-_{(aq)}$ so that

$$[Tl^+] = [OH^-] = 0{\cdot}170 \text{ mol dm}^{-3}$$

and the concentration of undissociated thallium(I) hydroxide is

$$0{\cdot}219 - 0{\cdot}170 = 0{\cdot}049 \text{ mol dm}^{-3}$$

The dissociation constant *K* of thallium(I) hydroxide can now be calculated.

$$K = \frac{[Tl^+][OH^-]}{[TlOH]} = \frac{0{\cdot}170^2}{0{\cdot}049} = 0{\cdot}590 \text{ mol dm}^{-3}$$

Other reactions which exhibit specific catalysis by either hydrogen or hydroxide ions can be used in a similar way. A comparison of the results for the acid hydrolysis of esters with strong acids and weak acids permits the calculation of the dissociation constants of the weak acids. The inversion of sucrose and the alkaline hydrolysis of esters have also been used in a similar way, but in the latter case the hydroxide ions are one of the reactants.

Experiment 5.3 Two investigations using a colorimeter

(*a*) *The kinetics of the iodination of acetone*

$$CH_3COCH_{3(aq)} + I_{2(aq)} = CH_3COCH_2I_{(aq)} + H^+_{(aq)} + I^-_{(aq)}$$

The reaction mixture has a red-brown colour initially, and as the iodine reacts the colour fades to pale yellow. A completely colourless solution is not obtained because the reaction reaches equilibrium.

The change in colour intensity can be followed using the circuit and apparatus shown in Figs. 5.2(a) and (b). The photocell† should be in circuit with a milli-

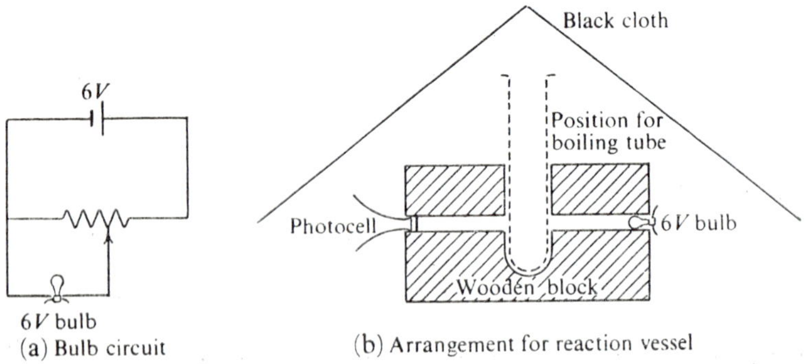

Fig. 5.2. Apparatus for use as a colorimeter

† Recommended type is the cadmium sulphide cell, obtainable from Proops Bros. Ltd, Tottenham Court Rd., London, W.1 (Order No. MKY 101-C).

ammeter reading to 1·0 mA and a battery giving about 2·4 V. Light of constant intensity is emitted by the lamp. Some is absorbed by the reaction mixture but some is transmitted and detected by the photocell. The reading on the milli-ammeter is a measure of the intensity of light transmitted by the reaction mixture. As the intensity of the colour decreases, the intensity of transmitted light rises, which in turn causes a rise in the current. A calibration plot is required to enable a conversion to be made from current in mA to concentration.

Quantities (for one run):

10 cm^3 acetone
10 cm^3 0·05 M iodine in 0·18 M potassium iodide solution
10 cm^3 2 M hydrochloric acid
10 cm^3 water.

Procedure

Quickly add the reactants to the boiling tube, shaking to ensure thorough mixing. Start a clock at the same time. Place the tube in the wooden block, and cover the whole of this part of the apparatus with a black cloth. Take the time for every 0·02 mA rise in the current until there is no further increase.

To obtain a calibration curve, make up a mixture in the same boiling tube containing 10 cm^3 acetone, 10 cm^3 of the iodine–iodide solution and 20 cm^3 water. This will have an iodine concentration equal to that of the initial reaction mixture. Place the tube in the wooden block and record the current. Then progressively dilute the mixture by known amounts using a solution containing one part of acetone to three parts of water, and record the concentration and the current in each case. Plot a graph of current against concentration. Using this calibration curve, construct a table of iodine concentrations at given times. Then plot iodine concentration against time on a fresh sheet of graph paper.

Discussion of results

The graph obtained is very similar to Fig. 4.9. Zero order kinetics are shown before the reaction achieves equilibrium. A detailed interpretation of this is given in Experiment 4.2 and in Chapter 6.

(*b*) *The kinetics of the reaction between bromide ions and concentrated nitric acid*

$$2Br^-_{(aq)} + 4HNO_{3(aq)} = Br_{2(aq)} + 2NO_{2(g)} + 2H_2O_{(l)} + 2NO^-_{3(aq)}$$

The reaction mixture is initially colourless, but as bromine is produced a yellow colour appears which gradually becomes orange-red.

Quantities: 15·0 cm^3 colourless concentrated nitric acid
15·0 cm^3 0·5 M potassium bromide solution.

Procedure

This is the same as in Experiment 5.3(*a*), but great care is needed because of the use of concentrated nitric acid. Retain the mixture after the reaction has been completed. To obtain a calibration curve, the completed reaction mixture should be diluted by known amounts of concentrated nitric acid. (CARE). Obtain a current reading for each known concentration. Using the calibration curve, calculate the concentration of bromine at the given time intervals, and then plot a concentration-time curve.

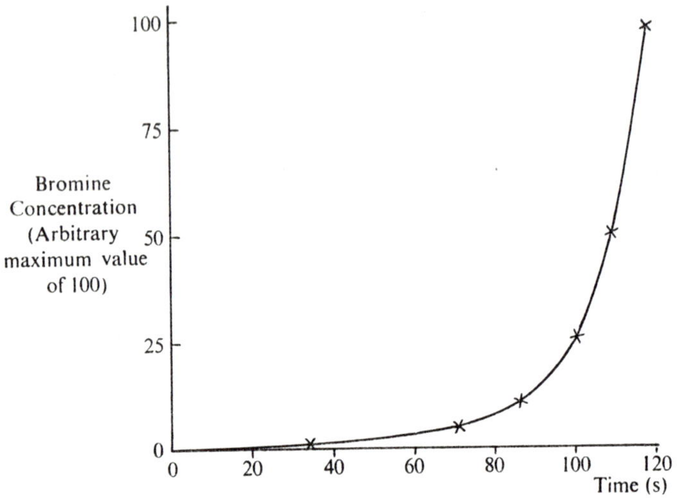

Fig. 5.3. A rate curve for the bromide-concentrated nitric acid reaction showing a marked autocatalytic effect

Discussion of results

Some typical results are plotted in Fig. 5.3. The rate of reaction is very low at first, but the reaction rapidly accelerates later on. The graph shows that the first 10 per cent of the reaction occurs in 88 s, but the last 90 per cent occurs in only 25 s. This is typical of an autocatalysed reaction (see page 113). Given that *yellow* concentrated nitric acid and bromide ions give a more or less normal rate curve with a high initial rate, what is likely to be the autocatalyst in this reaction?

B. The study of fast reactions

The reactions investigated in Chapter 4 and Experiments 5.1 to 5.3 all proceed at a rate measurable by conventional analytical techniques. The values of the half-life ($t_{1/2}$) for such reactions therefore must be greater than about 10^2 s. A reaction whose half-life is less than 10^2 s can be slowed down by lowering the reactant concentrations or the temperature. However, these modifications may still not produce a measurable rate if the reaction is a very fast one.

This section of the chapter deals with reactions which would be referred to as 'instantaneous' because they appear to go to completion immediately the reactants are mixed. The development of methods to study these fast and very fast reactions has shown that there are wide differences in their rates, but that *all* these rates are measurable. No reaction is instantaneous. The half-lives of these reactions vary from about 10^{-1} s to about 10^{-10} s.

The very fast reactions are all diffusion-controlled (see page 92), and the fastest known reaction is $H^+_{(aq)} + OH^-_{(aq)} = H_2O_{(l)}$, for which the second order rate constant k_2 is $1{\cdot}4 \times 10^{11}$ mol^{-1} dm^3 s^{-1} at 25°C. A rate constant of this magnitude gives a half-life of 10^{-11} s for 1 M concentrations (see page 83).

Two of the most important methods used in the study of these fast and very fast reactions are discussed below.

1. *Flow methods*

When two reactant solutions are mixed by pouring from one beaker to another the time of mixing is of the order of 1 second. In this time fast reactions would have gone to completion and no measurements on the rate could be made. If a method involving the mixing of the reactant solutions is to be used to study the rate of fast reactions then two conditions must be met:

1. the time of mixing must be shortened so that it is considerably less than the reaction time;
2. a way of making measurements must be found.

Flow techniques can meet these conditions for some reactions. The two reactant solutions are made to flow through two separate tubes into a mixing chamber. Under conditions of turbulence the mixing takes place in about 10^{-3} s. The reaction mixture then flows into an observation tube. With the flow of this solution at a constant rate, reaction occurs to a particular extent at each position along the tube. There is therefore no need for a method of analysis with a fast response. Each position corresponds to the lapse of a definite time interval (t s) after mixing given by

$$t = \frac{xa}{f} \qquad (6)$$

where x cm is the distance along the tube from the mixing chamber, a cm^2 is the cross-sectional area of the tube, and f cm^3 s^{-1} is the total flow rate. Some property of the reaction mixture, like colour intensity, conductance, etc., can be measured at various positions along the tube, and the corresponding reaction times calculated. Alternatively, the time taken for a given colour change can be found, and an example of this clock technique is described in Experiment 5.4.

Experiment 5.4 The flowing clock method†

Investigation of the kinetics of the Harcourt-Esson reaction.

$$2H^+_{(aq)} + H_2O_{2(aq)} + 2I^-_{(aq)} = 2H_2O_{(l)} + I_{2(aq)}$$

This reaction was studied in Experiment 4.5. An amount of monitor substance, corresponding to a small extent of the main reaction, was used. When the main reaction had produced sufficient iodine to react with all the monitor, the excess iodine turned the starch indicator blue. In Experiment 5.4 the same principle is used but the concentrations are raised so that the reaction time is about 10^{-1} s. The reactant solutions mix rapidly in tap B (see Fig. 5.4)

† See *J. Chem. Ed.*, 1963, **40**, 367.

and then pass along the flow tube. When all the monitor has reacted the starch turns blue and a colourless to blue boundary appears. The position of the boundary (x cm from the mixing chamber) is determined by the flow rate (f), the cross sectional area (a) of the tube and the reaction time (t) [see equation (6)]. If the amount of monitor is kept constant for each run then the effect of varying the concentration of each reactant on t can be investigated as in Experiment 4.5. If the amount of monitor is very small with respect to the amount of reactants then $1/t$ will be a good measure of the initial rate.

The flow tube has a length of 1 to 1·5 m and a bore of about 2 mm. Tap A is two-way and tap B three-way, and the latter must have a bore of about the same size as the flow tube. Pressure tubing should be used for all the attachments around the aspirator. The inlet tubes should have a diameter of at least 6 mm.

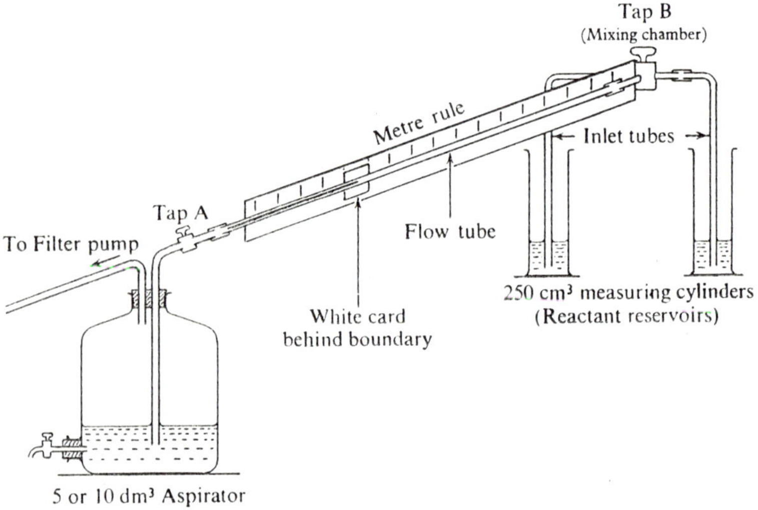

Fig. 5.4. The flowing clock apparatus

General comments

Turbulent flow, not streamlined flow, is essential. The above apparatus requires a total flow rate of at least 6 $cm^3\ s^{-1}$ (3 $cm^3\ s^{-1}$ from *each* reservoir) for turbulence. To detect whether the flow is turbulent or streamlined observe the blue boundary. If it is sharp and sensitive to small changes in water pressure then the flow is turbulent. A streamlined flow is indicated by an indistinct boundary which does not move when the water pressure is varied, and the appearance of blue 'streamers' at varying positions throughout the tube. If the flow is streamlined then a higher flow rate is required for turbulence and the most effective way of obtaining this is to use a filter pump on a tap connected to a water main with a higher pressure. The limiting conditions for turbulent flow in a given apparatus can be calculated. (See Appendix D.)

A fluctuating water pressure results in 'jumping' of the boundary. The aspirator helps to reduce the effects of fluctuations, but it may be necessary to choose a time for the experiment when little or no water is being used throughout the building.

(*a*) *Investigations at high hydrogen ion concentration* (~0·25 M)

Quantities and concentrations

'50 vol' hydrogen peroxide (dilute '100 vol' with starch solution)
Bench (1 M) sulphuric acid
0·1 M potassium iodide
Monitor: 0·01 M sodium thiosulphate (dilute 0·1 M solution with starch solution).

	Reservoir 1 cm³			*Reservoir* 2 cm³		
	Peroxide	*Acid*	*Water*	*Iodide*	*Thiosulphate*	*Starch*
Exp. 1	125	125	0	150	50·0	50·0
Exp. 2	62·5	125	62·5	150	50·0	50·0
Exp. 3	125	62·5	62·5	150	50·0	50·0
Exp. 4	125	125	0	75·0	50·0	125

Procedure

Turn on the mains water tap and reduce the pressure in the aspirator by keeping tap A closed. Cover the aspirator in case of implosion. Fill the two reservoirs with tap water. Open taps A and B so that water is drawn through the flow tube from both reservoirs. Adjust tap B so that the flow rates from the two reservoirs are about equal, and keep tap B at this adjustment for all succeeding runs.

Prepare the reservoir solutions in the above proportions. The solutions can be measured directly into the reservoir cylinders from stock solutions. Mix each solution *thoroughly*. Again reduce the pressure in the aspirator and then open tap A. Adjust the main water tap or tap A so that the blue boundary appears at a convenient position in the flow tube. Find this position accurately using a white card behind the boundary and measure its distance from tap B. Measure the flow rate from *each* cylinder by finding the time taken for the level to drop 50·0 cm³ in each. Calculate each flow rate in $cm^3 \ s^{-1}$ and sum them to give the total flow rate.

Wash out the apparatus, empty the aspirator, and repeat the procedure for each run. In each case, the distance, x, is required and the reaction time, t, which is calculated using equation (6) and under these conditions is the time for about 1 per cent of the reaction to be completed.

(*b*) *Investigations at low hydrogen ion concentration* (~0·0025 M)

Repeat the procedure of part (*a*) using the following quantities and concentrations.

'50 vol' hydrogen peroxide (dilute '100 vol' with water)
0·01 M sulphuric acid (dilute bench (1 M) acid with starch solution)
1 M potassium iodide
Monitor: 0·01 M sodium thiosulphate [prepared as in (a)].

	Reservoir 1 cm³			*Reservoir* 2 cm³		
	Peroxide	*Acid*	*Water*	*Iodide*	*Thiosulphate*	*Starch*
Exp. 5	125	125	0	100	50·0	100
Exp. 6	125	62·5	62·5	100	50·0	100

This investigation could be extended by varying the peroxide concentration and the iodide concentration as well as the hydrogen ion concentration.

Discussion of results

The following results (Table 5.3) were obtained using the above apparatus and concentrations.

Table 5.3

Experiment		*Total flow rate* f (cm³ s⁻¹)	*Distance of boundary from tap B* x (cm)	*Reaction time* t (s) $\left(t = \frac{x.a}{f} = \frac{x.0{\cdot}03}{f}\right)$
high [H⁺]	1	8·0	51	0·19
		7·6	46	0·18
	2	7·7	108	0·42
		7·3	96	0·39
		6·1	81	0·40
	3	7·9	90	0·34
	4	7·3	89	0·37
low [H⁺]	5	8·5	44	0·16
	6	8·5	47	0·17

Deduce the orders with respect to hydrogen peroxide, iodide ions, and hydrogen ions at high [H^+], and the order with respect to hydrogen ions at low [H^+]. If experiments were performed to find the orders with respect to hydrogen peroxide and iodide ions at low [H^+], deduce these as well. Experiments 1 and 2 show that the results are reproducible to within about 5 per cent. Can the orders be considered integers within this experimental error?

The results of Experiments 5 and 6 are striking because the amount of monitor used is such that 40 per cent of the hydrogen ions are used in Experiment 5 before the blue boundary appears and 80 per cent are used in Experiment 6. Yet the reaction time is still the same within experimental error.

These results are in accordance with the rate expression given in Chapter 6 for this reaction.

2. *Relaxation techniques*

These techniques are particularly valuable because they can be applied to some of the fastest reversible gaseous and solution reactions. Mixing is avoided so that the reaction times are not limited by the time of mixing.

There are a variety of methods but they all rely on the same principle. The reaction is allowed to reach equilibrium under controlled conditions. Then one of the factors which affects the equilibrium (e.g. temperature, pressure) is suddenly changed so that the reaction is no longer in a state of equilibrium. Reaction occurs and equilibrium is regained under the new conditions. The rate of reaction from the non-equilibrium state to the equilibrium state is measured by determining the rate of change of some suitable property which is related to the concentration of one of the reactants.

For fast reactions, the time taken for equilibrium to be regained is very short (about 10^{-4} s or less) and this imposes two very stringent requirements on the methods.

(*a*) The factor which is changed to disturb the equilibrium must be changed very rapidly. The time taken for the change must be of the order of, or shorter than, the half-life of the reaction.

(*b*) The detection device must respond very quickly to the changes which occur in the reaction mixture, and the response, too, has to be recorded quickly.

Factors which can be used to disturb an equilibrium are pressure, temperature, and intensity of an electric field. Two commonly used detecting methods involve the measurement of conductance and colour intensity.

Questions and exercises

1. Use the flowing clock technique to investigate the kinetics of the formaldehyde-hydrogen sulphite reaction. (See *J. Chem. Ed.*, 1963, **40**, 366).
2. Using the concentrations and procedure in Experiment 5.3(*a*), find the order with respect to hydrogen ions and acetone in the iodination of acetone.
3. In acid aqueous solution, diazoacetic ester decomposes as follows:

$$CHN_2.CO_2C_2H_{5(aq)} + H_2O_{(l)} = CH_2OH.CO_2C_2H_{5(aq)} + N_{2(g)}$$

The rate expression is:

$$-\frac{d[\text{ester}]}{dt} = k_2[H^+][\text{ester}]$$

Suggest a way of establishing this rate expression. Also calculate the dissociation constant of acetic acid from the following results.

Catalyst	$10^4 \times k_1$ min^{-1}
0·000909 M HNO_3	345
0·00182 M HNO_3	703
0·0182 M CH_3CO_2H	218

6 Order and mechanism

In this chapter and the next two chapters, an attempt is made to account for the experimental results obtained in Chapters 2, 3, 4 and 5. This involves the development of a theory about what is happening to reactant molecules before and during the reaction.

In its simplest form, a theory is a convenient way of picturing what is happening in the form of an analogy. This way of explaining the unfamiliar in terms of the familiar is often called a *model*. The model needs to explain the experimental observations qualitatively. When a suitable model has been devised, it may provide the basis for a mathematical treatment from which quantitative predictions can sometimes be made, and the model can then be tested against observations. This comparison may result in a need to modify the model to bring about close agreement between the predictions and the observations. When this has been achieved the model is a useful one and may attain the status of a theory.

This chapter begins with the development of a model of reaction by collision based on the Kinetic Theory model. The aim is to account for experimental rate expressions. In the next chapter, attempts are made to predict the magnitude of rates of reaction and their variation with temperature. It will be found that the model needs several alterations in order to obtain agreement between prediction and experiment.

An elementary model of reaction by collision for gases

It is easier to develop first a model for gaseous reactions rather than for reactions in solution, because the former are not complicated by the presence of large numbers of solvent molecules.

The Kinetic Theory of Gases pictures a gas as consisting of a very large number of molecules moving in random directions in straight lines. It is further assumed that there is a random distribution of velocities and therefore of translational kinetic energies. The molecules, which are regarded as hard spheres, undergo collisions, either with other gas molecules or with the walls of the container. In a mixture of two gases, A and B, there will be collisions between like molecules, unlike molecules and also between each kind of molecule and the walls of the container. If the two gases do not react together chemically, all of the collisions are assumed to be perfectly elastic and do not alter the types of molecule present. However, if the two gases do react chemically, it is reasonable to suppose that reaction can only occur when two reactant molecules collide.

Consider a mixture of two gases A and B which react according to the equation:

$$A_{(g)} + B_{(g)} = \text{products}$$

It is clear that, if the concentrations of either gas or both gases were increased, the frequency of collisions between A and B molecules would increase. This increase in collision frequency can be expected to increase the rate of reaction.

Some collisions are very energetic because they involve molecules with high kinetic energies, whereas others are less energetic. It may be that only the more energetic collisions are successful in forming products. This is discussed in Chapter 7 where it is shown that:

Rate of reaction $\propto$	number of successful $\propto$	total number of
	A/B collisions	A/B collisions
	(per unit volume,	(per unit volume
	per unit time)	per unit time)

Therefore an increase in the concentration of either gas will increase the rate of reaction. What does this model have to say about the *precise* dependence of rate on concentration? In other words, can the rate expression be predicted by applying the model to the stoichiometric equation?

The likelihood of one particular A molecule undergoing a collision with any B molecule in a given time depends on the number of B molecules per unit volume i.e. the molecular concentration of B. If this number per unit volume is doubled then the chance of a collision between the particular A molecule and any B molecule is doubled. This relationship can be expressed as:

Collision frequency for one A molecule $\propto$ [B]
(per unit volume, per unit time)

where [B] is the *molar* concentration of B.

However, there is more than one A molecule present. If there are 100 A molecules, then the likelihood of a collision between A and B molecules is 100 times greater than if there is only one A molecule present. The total collision frequency therefore depends on both [A] and [B] and

Collision frequency $\propto$ [A][B]
(per unit volume, per unit time)

Hence

Rate of reaction $\propto$ [A][B]
(mol per unit volume, per unit time)

For the reaction whose equation is

$$2A_{(g)} = \text{products},$$

the model predicts that

$$\text{Rate of reaction} \propto [A][A] \qquad \text{or } [A]^2$$

Similar arguments may be applied to reactions of other types. For instance, consider the reaction

$$4HBr_{(g)} + O_{2(g)} = 2H_2O_{(g)} + 2Br_{2(g)}$$

The equation may be rewritten as

$$HBr_{(g)} + HBr_{(g)} + HBr_{(g)} + HBr_{(g)} + O_{2(g)} = 2H_2O_{(g)} + 2Br_{2(g)}$$

so that the model predicts:

$$\text{Rate of reaction} = -\frac{d[O_2]}{dt} \propto [HBr]^4[O_2]$$

The model can be tested by comparing some predicted rate expressions with experimental rate expressions for some gaseous reactions (see Table 6.1). The table shows that only occasionally is there agreement between the model and experiment. Thus, this simple collision model cannot be applied generally, and it is necessary to modify it. One possible modification is well illustrated by referring to the hydrogen bromide–oxygen reaction.

Table 6.1. Rate expressions for some gaseous reactions

Reaction	*Rate expression predicted from the collision model*	*Experimental rate expression*
$H_2 + I_2 = 2HI$	Const. $[H_2][I_2]$	Const. $[H_2][I_2]$
$2HI = H_2 + I_2$	Const. $[HI]^2$	Const. $[HI]^2$
$H_2 + Br_2 = 2HBr$	Const. $[H_2][Br_2]$	$\dfrac{\text{Const. }[H_2][Br_2]^{1/2}}{1 + \text{Const.}\dfrac{[HBr]}{[Br_2]}}$
$H_2 + Cl_2 = 2HCl$	Const. $[H_2][Cl_2]$	Not yet completely solved
$4HBr + O_2 = 2H_2O + 2Br_2$	Const. $[HBr]^4[O_2]$	Const. $[HBr][O_2]$
$2N_2O_5 = 2N_2O_4 + O_2$	Const. $[N_2O_5]^2$	Const. $[N_2O_5]$
$2NO + O_2 = 2NO_2$	Const. $[NO]^2[O_2]$	Const. $[NO]^2[O_2]$
$CH_3CHO = CH_4 + CO$	Const. $[CH_3CHO]$	Const. $[CH_3CHO]^{3/2}$
$C_2H_4 + I_2 = C_2H_4I_2$	Const. $[C_2H_4][I_2]$	Const. $[C_2H_4][I_2]^{3/2}$

(See Chapter 10 for fuller discussion of some of these reactions.)

The simple model assumes that this reaction occurs by the collision of four hydrogen bromide molecules and one oxygen molecule. How likely is such a collision? It is less likely to occur than a collision involving a smaller number of molecules. The most likely collisions are those between two molecules, three-molecule collisions being less likely, and so on. In the absence of any contradictory evidence it is generally accepted that collisions involving more than three molecules are extremely unlikely and do not make any significant contribution to the rate of reaction. It must then be suggested that this reaction occurs by several simple steps instead of one complex collision. Each step in this *reaction mechanism* (or *pathway*) would involve collisions between two or three molecules instead of five. Only by postulating a mechanism with several simple collisions can a measurable rate be envisaged for this reaction.

It is usually necessary to postulate a mechanism where the rate expression predicted from the stoichiometric equation disagrees with the experimental expression. In the few cases where there is agreement, the reaction is usually *elementary*, i.e. proceeds in one step as indicated by the stoichiometric equation.† Reactions which are considered to proceed in a series of steps are called *complex*.

Any mechanism proposed for a complex reaction must be capable of accounting for the experimental rate expression. This does not establish the validity of the mechanism but it is a first, essential requirement. A possible mechanism for the hydrogen bromide–oxygen reaction will serve as an example.

† This is not always so. See Chapter 10 for a discussion of the hydrogen–iodine reaction.

Mechanism 1

$$\begin{aligned} HBr + O_2 &\rightarrow HOOBr && \text{Step 1} \\ HOOBr + HBr &\rightarrow 2HOBr && \text{Step 2} \\ \left.\begin{aligned} HOBr + HBr &\rightarrow Br_2 + H_2O \\ HOBr + HBr &\rightarrow Br_2 + H_2O \end{aligned}\right\} &&& \text{Step 3} \end{aligned}$$

Summing these steps produces the stoichiometric equation for the reaction.

$$4HBr_{(g)} + O_{2(g)} = 2Br_{2(g)} + 2H_2O_{(g)}$$

Since each step is not made up of any simpler steps (i.e. is elementary), it is possible to apply the simple collision model to each one. Therefore:
Rate of removal of hydrogen bromide (or oxygen) is:

$$-\frac{d[HBr]}{dt} = \text{Const. } [HBr][O_2] \qquad \text{(for Step 1)}$$

$$-\frac{d[HBr]}{dt} = \text{Const. } [HOOBr][HBr] \qquad \text{(for Step 2)}$$

$$-\frac{d[HBr]}{dt} = \text{Const. } [HOBr][HBr] \qquad \text{(for Step 3)}$$

These rates are the predicted rates for each individual step in the mechanism. How can they be used to account for the observed rate of the whole reaction?

It is assumed that for most reaction mechanisms one of the steps is slower than the others, and if this is the first step it is known as the *rate determining step*. The rate of reaction cannot be higher than the rate of the slowest step. (Chain reactions provide an exception to this.)

An analogy is the rate of entry of spectators through a turnstile into a football ground. There are two main processes which could affect this rate:

1. The rate at which people arrive in the queue.
2. The rate at which people pass through the turnstile.

On a busy day, when the queue builds up quickly, the rate of entry would be solely determined by the speed of operation of the turnstile. Any fluctuations in the rate of queue-building would be irrelevant to the rate at which people enter the ground. The rate determining step is therefore the rate at which people pass through the turnstile. Although there are two processes, only one is rate determining.

Which step in the mechanism for the hydrogen bromide–oxygen reaction must be the slowest if the predicted rate expression is to agree with the experimental rate expression? Clearly this must be step 1, since

$$\text{Rate of step 1} = -\frac{d[HBr]}{dt} = \text{Const. } [HBr][O_2]$$

which is also the experimental rate expression. It must therefore be postulated that steps 2 and 3 are fast compared with step 1.

These arguments do not *prove* that the postulated mechanism is correct. They merely show it to be possible if it is assumed that step 1 is rate determining.

Another mechanism consistent with the stoichiometry of this reaction is:

Mechanism 2

$$2HBr + O_2 \rightarrow 2HOBr \qquad \text{Step 1}$$

$$\left.\begin{array}{l} HOBr + HBr \rightarrow Br_2 + H_2O \\ HOBr + HBr \rightarrow Br_2 + H_2O \end{array}\right\} \qquad \text{Step 2}$$

If step 1 were the slow step, the rate of reaction would be predicted as

$$-\frac{d[HBr]}{dt} = \text{Const. } [HBr]^2 [O_2]$$

and if step 2 were the slow step, it would be

$$-\frac{d[HBr]}{dt} = \text{Const. } [HOBr][HBr]$$

Neither prediction resembles the experimental rate expression (even if, in the second expression, [HOBr] was written in terms of [HBr] and $[O_2]$) so that mechanism 2 must be discarded.

Order and molecularity

The number of species involved in each step is known as the *molecularity* of that step. All steps in the first mechanism for the hydrogen bromide–oxygen reaction involve two molecules and are called *bimolecular*. In the second mechanism, step 1 is *termolecular* (i.e. involves three molecules) whilst step 2 is again bimolecular.

Molecularity should not be equated with the order of a reaction. The former refers to a *theoretical* mechanism whereas the latter is *experimentally* determined. In the hydrogen bromide–oxygen reaction, because the first step in mechanism 1 is rate determining, both the order and the molecularity of the reaction are 2. Order and molecularity are not always equal numerically (see the reaction between bromide and bromate ions in acid solution discussed below).

Reactions in solution

There are difficulties about applying the above model for gaseous reactions to reactions in solution. The reacting species in solution are commonly ions in an ionizing solvent whereas in gases they are neutral molecules. Moreover, the presence of very large numbers of solvent molecules in solution complicates the situation. Solute particles are surrounded closely by solvent molecules and it would be expected that many collisions would occur between the particles and the solvent molecules as well as between the particles themselves. Since there will always be a far higher concentration of solvent molecules than reactant species, the effects of the solvent are clearly very important in determining the motion of the reactant species and the nature of their collisions.

Despite these complications, the fundamental assumption that reaction can only occur by the collision of reactant species remains as the basis for the model. How do its predictions of rate expressions for reactions in solution compare with the experimental expressions determined in Chapters 3, 4, and 5? Table 6.2 is a complete list of these.

Table 6.2. Some rate expressions for reactions in solution

Stoichiometric equation (*for convenience the symbol* (aq) *is omitted*)	*Predicted rate expression*	*Experimental rate expression*	*References*
$S_2O_3^{2-} + 2H^+ = S + SO_2 + H_2O$	Const. $[S_2O_3^{2-}][H^+]^2$	Const. $[S_2O_3^{2-}] . \frac{[H^+]}{1 + \text{Const.} [H^+]}$	Experiments 3.2 and 3.3 Appendix A
$2H_2O_2 = 2H_2O + O_2$ (MnO_2 catalyst)	Const. $[H_2O_2]^2$	Const. $[H_2O_2]$	Experiment 3.4
$C_6H_5N_2^+ + H_2O = C_6H_5OH + N_2 + H^+$	Const. $[C_6H_5N_2^+][H_2O]$	Const. $[C_6H_5N_2^+]$ (a term in $[H_2O]$ may also be present)	Experiment 4.1 and page 76
$2I^- + S_2O_8^{2-} = I_2 + 2SO_4^{2-}$	Const. $[I^-]^2[S_2O_8^{2-}]$	Const. $[I^-][S_2O_8^{2-}]$	Experiment 4.3 and Question 4, Chapter 4
$CH_3COCH_3 + I_2 = CH_3COCH_2I + H^+ + I^-$ (in acid solution)	Const. $[CH_3COCH_3][I_2]$	Const. $[CH_3COCH_3][H^+]$	Experiment 4.2
$6H^+ + BrO_3^- + 5Br^- = 3Br_2 + 3H_2O$	Const. $[H^+]^6[BrO_3^-][Br^-]^5$	Const. $[H^+]^2[BrO_3^-][Br^-]$	Experiment 4.4
$H_2O_2 + 2H^+ + 2I^- = 2H_2O + I_2$	Const. $[H_2O_2][H^+]^2[I^-]^2$	Const. $[H_2O_2][I^-]$ + Const. $[H_2O_2][I^-][H^+]$	Experiments 4.5 and 5.4

Notes: 1. In some experiments, the orders are not exactly integral. These have been taken to the nearest integer in this table.

2. Each rate expression may differ according to the reactant (or product) used to express the rate.

e.g. $-d[BrO_3^-]/dt = -\frac{1}{5} d[Br^-]/dt = -\frac{1}{6} d[H^+]/dt = \text{Const.} [H^+]^2[BrO_3^-][Br^-]$

Again it seems that there is very little agreement between the collision model and experiment. Indeed, in the iodination of acetone, there is the surprising case of a rate expression which does not contain the concentration of one of the reactants (i.e. iodine). Therefore, the reaction cannot be a simple collision between an iodine molecule and an acetone molecule. Thus it is necessary to postulate that many reactions in solution occur in a series of simple steps rather than in one step (i.e. are complex), and that one of these steps is rate determining.

Possible mechanisms for some reactions in solution

The Harcourt-Esson Reaction

$$H_2O_{2(aq)} + 2H^+_{(aq)} + 2I^-_{(aq)} = 2H_2O_{(l)} + I_{2(aq)}$$

In Experiment 4.5, the acid concentration was high and effectively constant throughout, and the dependence of the rate on the hydrogen ion concentration was not investigated. But the results of Experiment 5.4 show that at low acid concentrations the experimental rate expression is the same as that found in Experiment 4.5. (This is referred to again on page 78.)

This rate expression is:

$$-\frac{d[H_2O_2]}{dt} = \text{Const. } [H_2O_2][I^-]$$

If the reaction were elementary the predicted rate expression would be:

$$-\frac{d[H_2O_2]}{dt} = \text{Const. } [H_2O_2][H^+]^2[I^-]^2$$

since the stoichiometric equation involves one hydrogen peroxide molecule, two hydrogen ions and two iodide ions. The reaction must therefore be complex, and three mechanisms are given below, all of which involve several simple steps. The steps in each mechanism can be summed to give the stoichiometric equation. The essential question is whether they can be shown to be in agreement with the experimental rate expression.

Mechanism 1	$H_2O_2 + I^- \rightarrow H_2O + OI^-$	Step 1
	$H^+ + OI^- \rightarrow HOI$	Step 2
	$HOI + H^+ + I^- \rightarrow I_2 + H_2O$	Step 3
Mechanism 2	$H_2O_2 + H^+ + I^- \rightarrow H_2O + HOI$	Step 1
	$HOI + H^+ + I^- \rightarrow H_2O + I_2$	Step 2
Mechanism 3	$H_2O_2 + H^+ + 2I^- \rightarrow H_2O + OI^- + HI$	Step 1
	$OI^- + HI + H^+ \rightarrow H_2O + I_2$	Step 2

The rate expression for each step in each mechanism can be predicted using the simple collision model. One step in each mechanism is assumed to be slower than the others. The only step which has a rate expression identical to the experimental rate expression is step 1 of mechanism 1 where:

$$-\frac{d[H_2O_2]}{dt} = \text{Const. } [H_2O_2][I^-]$$

Hence mechanism 1 is a possibility if it is assumed that step 1 is rate determining. Mechanisms 2 and 3 are discarded because they cannot provide agreement between prediction and experiment.

From the discussions of the hydrogen bromide–oxygen reaction and the Harcourt-Esson reaction, it should now be clear that the experimental rate expression provides information about the possible nature of the slow step in the mechanism. Having decided on a possible slow step, the mechanism has to be completed by adding other steps which are presumed to be fast relative to the slow step. The sum of the steps must give the stoichiometric equation.

For mechanism 1, it was necessary to postulate that step 1 was the slow step in order that the mechanism agreed with the experimental rate expression. This step is the one which determines the overall rate of reaction. However, it is frequently necessary to postulate a mechanism for some reactions in which the slow step is not the first step. In these cases it is more difficult to show how the mechanism agrees with the experimental rate expression; and the slow step is not the only step which contributes to the overall rate. An example of this is discussed on page 77, but for a more detailed, general treatment the reader should consult Appendix F.

Experimental rate expressions provide important evidence which enables possible mechanisms to be postulated, but it must be emphasised that a *complete* mechanism requires much more evidence before it is considered acceptable. Direct proof of the existence of an intermediate species (i.e. a species appearing in the mechanism but not in the stoichiometric equation) is quite rare because they are so short-lived. In many cases all that can be decided is whether the mechanism is 'reasonable': are the intermediate species likely to exist; have they been identified in another, analogous reaction; is one bond more likely to be broken than another; are the individual steps bimolecular or termolecular; do the steps involve ions with unlike or like charges? The answers to these questions are particularly important when there is more than one possible mechanism consistent with the experimental rate expression. It must be admitted that the evidence for and against many mechanisms is rather flimsy and none must be accepted without a great deal of experimental evidence.

The iodination of acetone in acid solution (Experiment 4.2)

$$CH_3COCH_{3(aq)} + I_{2(aq)} = CH_3COCH_2I_{(aq)} + H^+_{(aq)} + I^-_{(aq)}$$

For any one run the rate expression is:

$$-\frac{d[CH_3COCH_3]}{dt} = \text{Const.}\ [CH_3COCH_3]$$

(The reaction is zero order with respect to I_2.) However, the rate constant is proportional to the initial hydrogen ion concentration so that the complete rate expression is:

$$-\frac{d[CH_3COCH_3]}{dt} = \text{Const.}\ [CH_3COCH_3][H^+]$$

It could be that the first, and rate determining, step is

$$CH_3COCH_3 + H^+ \rightarrow \text{intermediate} \qquad \text{Step 1 (slow)}$$

In order to satisfy the stoichiometry of the reaction, a second step, which would be fast compared to step 1, might be

$$\text{intermediate} + I_2 \rightarrow CH_3COCH_2I + 2H^+ + I^- \qquad \text{Step 2 (fast)}$$

The predicted reaction rate would then be equal to the rate of the slower step and given by

$$-\frac{d[CH_3COCH_3]}{dt} = \text{Const. } [CH_3COCH_3][H^+]$$

This mechanism is consistent with the result that the rate of reaction is independent of the iodine concentration. Iodine only appears in the fast second step which has no influence on the overall rate of reaction. A reference to the turnstile analogy should clarify this point. On a busy day, the rate of entry into the ground is determined only by the efficiency of the turnstile operator. This holds true if the queue contains 100 people or 1000 people. Similarly iodine can be added to the reaction mixture in large or small quantities without affecting the rate of reaction, which is determined solely by step 1.

The decomposition of benzene diazonium chloride in aqueous solution (Experiment 4.1)

$$C_6H_5N^+_{2(aq)} + H_2O_{(l)} = C_6H_5OH_{(aq)} + N_{2(g)} + H^+_{(aq)}$$

The decomposition of benzene diazonium chloride shows first order kinetics. This suggests that the rate expression is:

$$-\frac{d[C_6H_5N_2^+]}{dt} = \text{Const. } [C_6H_5N_2^+]$$

However, the concentration of water was effectively constant in each run, so that it is not known whether its variation affects the rate of the reaction. A second order rate expression is therefore a possibility:

$$-\frac{d[C_6H_5N_2^+]}{dt} = \text{Const. } [C_6H_5N_2^+][H_2O]$$

If the reaction is second order, then a possible mechanism would be the collision of a water molecule with the diazonium cation as indicated by the stoichiometric equation. However, if it could be shown that the reaction is first order in the diazonium cation but zero order in water (and in any other species present), it would be necessary to postulate a unimolecular rate determining step such as:

$$C_6H_5N_2^+ \rightarrow C_6H_5^+ + N_2$$

The phenyl ion would then react with water.

$$C_6H_5^+ + H_2O \rightarrow C_6H_5OH + H^+$$

The bromate–bromide reaction in acidic solution (Experiment 4.4)

In all the above mechanisms it was necessary to assume a first step which was rate determining; and in these cases the species, and their proportions concerned in this slow step, were indicated by the rate expression. However, for some reactions a slow first step is most unlikely, and a slow succeeding step must be postulated.†

The experimental rate expression determined in Experiment 4.4 was:

$$-\frac{d[BrO_3^-]}{dt} = \text{Const. } [H^+]^2[BrO_3^-][Br^-]$$

If the first step were rate determining it would be:

$$2H^+ + BrO_3^- + Br^- \rightarrow \text{intermediates}$$

A molecularity of four demands a simultaneous collision of four ions. This is far less likely to occur than the collision of two ions. A more likely mechanism would be one involving several bimolecular steps as given below.

In this possible mechanism some rapid steps occur prior to the slow step. For example,

$$H^+ + Br^- \rightleftharpoons HBr \qquad \text{Steps 1}a \text{ and 1}b \text{ (fast)}$$

$$H^+ + BrO_3^- \rightleftharpoons HBrO_3$$

These steps would have time to come to equilibrium because of the slow succeeding step

$$HBr + HBrO_3 \rightarrow HOBr + HOOBr \qquad \text{Step 2 (slow)}$$

The two intermediates have been postulated because they are known to exist in other reactions and because they can then react further in two bimolecular fast steps to give the products.

$$HOOBr + HBr \rightarrow 2HOBr \qquad \text{Step 3 (fast)}$$

$$HOBr + HBr \rightarrow H_2O + Br_2 \qquad \text{Step 4 (fast)}$$

The mechanism can be shown to agree with the stoichiometry of the reaction by multiplying step 4 by three and summing all the steps.

It should be noted that the slow step does not involve, in this case, the actual species either in the stoichiometric equation or in the rate expression. The predicted rate of reaction is given as Const. $[HBr][HBrO_3]$ since this is the rate expression for the slow step, and yet the experimental rate expression involves hydrogen ions, bromide ions and bromate ions.

The stoichiometric equation is best written in ionic form because HBr and $HBrO_3$ are strong acids which exist in dilute solution as ions. If they were weak acids, the best stoichiometric equation would be

$$5HBr + HBrO_3 = 3Br_2 + 3H_2O$$

How is it possible, then, for the species HBr and $HBrO_3$ to be involved in the mechanism? In a dilute solution containing hydrogen ions, bromide ions and

† In these cases, the concept of one rate determining step is not really valid. For a discussion of this, see Appendix F.

bromate ions, there will be collisions between pairs of ions. Only in the case of the ions with unlike charges will these collisions bring the ions near enough for them to be held together by electrostatic attraction. The species thus formed are best represented as H^+Br^- and $H^+BrO_3^-$ and are known as *ion-pairs*. Subsequent collisions of each of these ion-pairs with water molecules may be vigorous enough to separate them. Alternatively, an H^+Br^- ion-pair may collide with a $H^+BrO_3^-$ ion-pair and may react according to step 2 of the mechanism. Even if there were only a few such ion-pairs in the reaction mixture their reaction together would cause the step 1 equilibria to be displaced to the right hand side. This would produce a few more ion-pairs for step 2 to occur, and so on.

It should now be seen that the rate expressions involving $[HBr][HBrO_3]$ and $[H^+]^2[Br^-][BrO_3^-]$ are equivalent. Thus the assumptions that step 2 is the slow step and that the forward and reverse reactions in step 1 are fast are in agreement with experimental observation, and the above mechanism can be considered as a possibility for this reaction.

In this mechanism the *molecularity* of the slow step is two whereas the total *order* is four. There is now a clear difference between the two terms.

Competing mechanisms

The football ground analogy can be used to explain the idea of competing mechanisms. Consider two turnstiles, one leading into a covered stand and one into an uncovered part of the ground. The rate of entry into the ground is now the sum of the two separate rates of entry through each turnstile. In bad weather it may be that most of this total rate would be accounted for by the rate of entry into the covered part of the ground and this 'pathway' is therefore used predominantly. The contribution to the total rate of entry at each turnstile might thus depend on the weather conditions.

The Harcourt-Esson reaction (Experiments 4.5 and 5.4) has a rate expression which varies with the experimental conditions. For instance, at low $[H^+]$ the rate expression is

$$-\frac{d[H_2O_2]}{dt} = \text{Const.}[H_2O_2][I^-]$$

whereas at higher $[H^+]$ (greater than about 0·066 M) the rate expression is nearer

$$-\frac{d[H_2O_2]}{dt} = \text{Const.}[H_2O_2][I^-][H^+]$$

At intermediate $[H^+]$ the order with respect to hydrogen ions is fractional and always between 0 and 1. It is found that the results are best explained by a rate expression which is the sum of two terms:

$$-\frac{d[H_2O_2]}{dt} = \text{Const.}[H_2O_2][I^-] + \text{Const.}[H_2O_2][I^-][H^+]$$

This is interpreted as resulting from two parallel, competing mechanisms which both agree with the stoichiometry of the reaction. One mechanism is predominant under one range of conditions and the other under a different range of conditions.

Each mechanism has its own, different rate determining step. For the $[H^+]$-dependent rate expression a possible first, and rate determining, step would be

$$H_2O_2 + I^- + H^+ \rightarrow \text{intermediates} \qquad \text{(slow)}$$

i.e. a termolecular step. The species produced by this step would then be converted to the products by a series of fast steps. The possible mechanism for the $[H^+]$-independent pathway has been discussed already (see page 74).

The existence of intermediates

It has already been mentioned that direct proof of the existence of an intermediate is not very common. Spectroscopic studies and also the use of radioactive tracers can sometimes provide direct proof. Ordinary analytical techniques are not applicable because these species are too unstable.

In a few cases, however, an intermediate is stable enough to be identified and isolated. In Experiment 6.1 an intermediate can be seen clearly in the reaction mixture and is relatively long-lived compared with most. Further examples are described in Experiments 8.1 and 8.2.

Experiment 6.1 Investigation of the intermediates in the reduction of iron(III) ions by thiosulphate ions

$$2Fe^{3+}_{(aq)} + 2S_2O^{2-}_{3(aq)} = 2Fe^{2+}_{(aq)} + S_4O^{2-}_{6(aq)}$$

Procedure

Dissolve about 5 g of sodium thiosulphate (hydrated) in 20 cm^3 of water and add 20 cm^3 of bench iron(III) chloride solution (about 0·5 M). Shake to mix the reactants. Observe the colour change. When the reaction is complete, test for the presence of iron(II) ions using potassium hexacyanoferrate(III) solution. On a separate sample of solution show that all the iron(III) ions have been consumed using potassium thiocyanate solution.

Discussion of results

The immediate appearance of a deep violet colour is probably due to the very rapid formation of the intermediate $Fe^{3+}S_2O_3^{2-}$.

$$Fe^{3+} + S_2O_3^{2-} \rightleftharpoons Fe^{3+}S_2O_3^{2-} \qquad \text{Step 1 (fast)}$$

The species $Fe^{3+}S_2O_3^{2-}$ is another example of an ion-pair (see page 78). Step 1 is a likely first step because oppositely charged ions are likely to associate if they can approach each other closely enough during collision.

The gradual fading of the colour would then be due to the decomposition of these ion-pairs in one or more steps. Because the fading of the colour is much slower than its appearance, the slow step is presumably one of these later steps.

More information about the nature of this slow step could be gained by investigating the kinetics of the reaction. The disappearance of the colour could be timed for different initial concentrations of iron(III) ions and thiosulphate ions. The resulting rate expression may indicate the *proportions* of iron(III) ions and thiosulphate ions involved in the slow step.

The reaction between iron(III) chloride solution and sulphur dioxide at room temperature is similar to that between iron(III) ions and thiosulphate ions; a red coloured intermediate is formed (possibly $Fe^{3+}SO_3^{2-}$) which reacts to form the products.

Summary

1. Reactions usually proceed in a series of elementary steps. Such *complex* reactions have rate expressions which cannot be predicted from the stoichiometric equation.
2. Some reactions are *elementary*, i.e. proceed in one step as represented by the stoichiometric equation. The rate expressions of elementary reactions can be predicted from the stoichiometric equation.
3. The collision model can only be applied to (*a*) elementary steps in the mechanism of a complex reaction, or to (*b*) an elementary reaction.
4. There is no *a priori* way of knowing whether a given reaction will be complex or elementary. Only experiment can decide this. Generally, complex stoichiometric equations indicate that the reaction is complex, but simple stoichiometric equations often accompany complex rate expressions.

Questions

1. In the Harcourt-Esson reaction the rate expression is:

$$-\frac{d[H_2O_2]}{dt} = k'[H_2O_2][I^-] + k''[H_2O_2][I^-][H^+]$$

and this is interpreted best in terms of competition between two pathways.
At 25°C, experimental results have shown that

$$k' = 1{\cdot}15 \times 10^{-2}\ \text{mol}^{-1}\ \text{dm}^3\ \text{s}^{-1}$$

and

$$k'' = 1{\cdot}75 \times 10^{-1}\ \text{mol}^{-2}\ \text{dm}^6\ \text{s}^{-1}$$

Calculate the $[H^+]$ when each pathway contributes to 50 per cent of the reaction.
Why is the concept of total order of a reaction not readily applicable to this reaction? What are the minimum and maximum total orders for the reaction, and under what conditions will they be observed?

2.

$$I_{2(aq)} + H_3PO_{2(aq)} + H_2O_{(l)} = H_3PO_{3(aq)} + 2I^-_{(aq)}$$

The rate expression for this reaction is

$$-\frac{d[H_3PO_2]}{dt} = k_1[H_3PO_2]$$

The rate is independent of the iodine concentration. How can this be interpreted? In view of the fact that this reaction takes place in aqueous solution, and that H_2O is a reactant in the equation, what is another possible rate expression?

3. The reduction of potassium iodate to potassium iodide by sulphurous acid is represented by the stoichiometric equation

$$KIO_{3(aq)} + 3H_2SO_{3(aq)} = KI_{(aq)} + 3H_2SO_{4(aq)}$$

The rate expression is:

$$-\frac{d[KIO_3]}{dt} = -\frac{1}{3}\frac{d[H_2SO_3]}{dt} = k_2[KIO_3][H_2SO_3]$$

Assuming that the first step in the mechanism is rate determining, write out a possible mechanism for this complex reaction. (The mechanism of the bromate–bromide reaction in acid solution may indicate some possible intermediates for the above reaction.)

7 Theories of reaction rates

This chapter begins by developing a detailed collision theory for reaction kinetics, and then continues by discussing some of the shortcomings of the theory. Finally, another theory is proposed which emphasizes what may happen between the collision of reactant molecules and the separation of product molecules.

A more detailed collision theory

The comparison of predictions from the model of reaction by collision and the experimental results may be continued and made more rigorous by attempting to calculate the theoretical rate at which reactions should occur if the assumptions underlying the theory are valid.

A mathematical treatment of the Kinetic Theory of Gases allows the calculation of the frequency of collisions between gas molecules. For a mixture of two gases A and B, the frequency of A/B collisions is given by:

$$Z = Yn_A n_B \qquad (1)$$

where Z is the number of collisions per dm^3 per second, n_A and n_B are the number of A and B molecules respectively per dm^3, and Y is a complicated term which involves the radii of A and B molecules, the Gas Constant, the molecular weights of A and B, and the absolute temperature (see Appendix C).

If the reaction whose stoichiometric equation is

$$A_{(g)} + B_{(g)} = \text{products}$$

takes place by the collision of A and B molecules, and not by a more complex mechanism then:

$$\text{Rate of reaction} = -\frac{d[A]}{dt} = -\frac{d[B]}{dt} \propto Z$$

If every collision produces a reaction then:

$$-\frac{d[A]}{dt} = Z \text{ molecules dm}^{-3}\text{ s}^{-1}$$

$$= \frac{Z}{N} \text{ mol dm}^{-3}\text{ s}^{-1} \qquad (2)$$

where N is the Avogadro Number.

Thus from (1) and (2),

$$\text{Rate of reaction} = \frac{Z}{N} = \frac{Yn_A n_B}{N} \text{ mol dm}^{-3}\text{ s}^{-1}$$

$$= YN.\frac{n_A}{N}.\frac{n_B}{N} \text{ mol dm}^{-3}\text{ s}^{-1}$$

$$= YN[A][B] \text{ mol dm}^{-3}\text{ s}^{-1} \qquad (3)$$

where [A] and [B] are now molar concentrations. The term YN in equation (3) should now be seen as the proportionality constant for the expression

$$\text{Rate of reaction} \propto [\text{A}][\text{B}]$$

This was developed in Chapter 6 for a bimolecular reaction. The *experimental* rate expression for this bimolecular reaction is:

$$-\frac{d[\text{A}]}{dt} = k_2[\text{A}][\text{B}]$$

where k_2 is the second order rate constant.

The prediction is made therefore that:

$$k_2 = YN \qquad (4)$$

The limitations which apply to this prediction should be noted carefully:

(*a*) The reaction considered is bimolecular. It could be either an elementary reaction or a single step in a more complex reaction. The above considerations, therefore, do not apply to unimolecular or termolecular processes.

(*b*) Equation (4) applies only if all collisions are successful. If, for some reason, only one in every three collisions caused a reaction, a factor of $\frac{1}{3}$ would be needed, i.e.

$$k_2 = \tfrac{1}{3}.YN$$

(*c*) The model used was for gaseous reactions, not for reactions in solution.

A comparison of the model with experimental results

Is equation (4) valid? It can be tested by comparing experimental values of k_2 for reactions which are known to be bimolecular with the value of k_2 predicted from this equation. Firstly a value for Y must be calculated.

When the temperature is 500 K, the sum of the radii of A and B molecules equal to 3×10^{-10} m, and the molecular weights of A and B equal to 10 and 100 the value of Y is about 10^{-13} dm^3 s^{-1} (see Appendix C for the calculation). Hence

$$\begin{aligned} YN &\sim 10^{-13} \times 10^{24} \\ &= 10^{11}\ \text{mol}^{-1}\ \text{dm}^3\ \text{s}^{-1} \end{aligned}$$

When the temperature is halved or doubled the value of Y, and hence YN, is only altered by a factor of $\sqrt{2}$ so that temperature makes very little difference and k_2 can be taken as about 10^{11} mol^{-1} dm^3 s^{-1} for a wide range of temperatures. Moreover, the result of the calculation does not vary a great deal if other reasonable values for the radii and the molecular weights are used. It is clear, therefore, that this predicted value of k_2 should vary little from reaction to reaction.

A rate constant of about 10^{11} mol^{-1} dm^3 s^{-1} is very large indeed, and its magnitude can be realized by calculating the corresponding half-life of the reaction. For unit concentrations of A and B, and using this predicted value for k_2,

$$\begin{aligned} t_{1/2} &= \frac{1}{c_0 k_2} \qquad \text{[Equation (16), Chapter 4]} \\ &= \frac{1}{1 \times 10^{11}} = 10^{-11}\ \text{s} \end{aligned}$$

Thus ten half-lives are 10^{-10} s, and during this time the concentrations of A and B would decrease from 1 mol dm^{-3} to $1 \times (\frac{1}{2})^{10} \sim 10^{-3}$ mol dm^{-3}. The reaction is so rapid that 99·9 per cent of the reactants are consumed in 10^{-10} s. It should be clear that many gaseous reactions cannot proceed at this high rate. A great number of such reactions have rates which are slow enough to enable simple measurements to be made on them. However, it is instructive to compare the predicted value of k_2 with a selection of experimental rate constants.

Care must be taken in the choice of rate constants. The comparison must be made, not with the overall rate constants for complex reactions, but either with rate constants for individual *bimolecular* steps in a complex reaction, or with the rate constants of reactions which occur by a single *bimolecular* step. This at once raises the problem that there are few reactions which are thought to be elementary, and it is necessary therefore to use results for reactions which will be unfamiliar to some readers of this book. Table 7.1 is a list of k_2 values for some gaseous reactions.

Table 7.1. Second order rate constants for elementary gaseous reactions

Reaction	k_2 *at* 773 K mol^{-1} dm^3 s^{-1}
$H_2 + I = H + HI$	90
$H + Br_2 = Br + HBr$	$8{\cdot}3 \times 10^{10}$
$Br + H_2 = H + HBr$	6×10^{5}
$HBr + Br = H + Br_2$	6×10^{-2}
$Cl + H_2 = HCl + H$	3×10^{9}
$2HI = H_2 + I_2$	$1{\cdot}2 \times 10^{-2}$
$D + H_2 = H + HD$	10^{9} (at 1 000 K)
$COCl + Cl = CO + Cl_2$	2×10^{11}
$NOCl + Cl = NO + Cl_2$	5×10^{9}
$2NOCl = 2NO + Cl_2$	10^{3}
$CH_3 + CH_3 = C_2H_6$	2×10^{10}
$CH_4 + Br = CH_3 + HBr$	5×10^{5}
$2\,CH_2{=}CH{-}CH{=}CH_2$ = 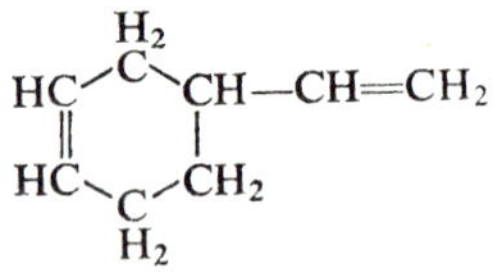	4

In Table 7.1, there is an extremely large variation in values for k_2 mainly towards very much lower values than 10^{11} mol^{-1} dm^3 s^{-1}. However, two reactions do have a value for k_2 in good agreement with this predicted value, and no reaction has a rate constant *greater* than 10^{11} mol^{-1} dm^3 s^{-1}. The model predicts well the *maximum* observed values for gaseous reactions.

It has already been mentioned that $k_2 = YN$ [Equation (4)] is only valid when it is assumed that all collisions between reactant species are successful in promoting a reaction. Clearly there are some reactions where this assumption seems justified, but for the majority of the reactions $k_2 \ll YN$, suggesting that only a fraction of the total collisions are successful. This fraction must be different for each reaction. In many cases the fraction would have to be extremely small in order to account for the experimental rate constants, and it is difficult to imagine at this stage why this should be.

Example

The reaction of HBr with Br is a bimolecular reaction with an observed rate constant of 6×10^{-2} mol^{-1} dm^3 s^{-1} at 500°C. The fraction of collisions which are successful must therefore be

$$\frac{6 \times 10^{-2}}{10^{11}} \sim 10^{-12}$$

The discrepancy between the observed k_2 and the predicted k_2 can therefore be accounted for by assuming that only one collision in about 10^{12} is successful.

Reactions in solution

Table 7.2 is a list of k_2 values for some reactions in solution. The assumptions on which k_2 was calculated to be about 10^{11} mol^{-1} dm^3 s^{-1} were based on a gaseous model of reaction by collision. It is obviously not justifiable to consider this as a predicted value for k_2 for reactions in solution without a detailed consideration of the motion of particles in solution. The complicating effects of the solvent (see page 72) make the calculation of the collision frequency of solute particles much more difficult. However, it is interesting to compare values for k_2 in Table 7.2 with those in Table 7.1.

Table 7.2. Second order rate constants for elementary reactions in solution.

Reaction (*for convenience, the symbol* (aq) *has been omitted*)	k_2 (*at about* 25°C) mol^{-1} dm^3 s^{-1}
$H^+ + SO_4^{2-} = HSO_4^-$	4×10^{10}
$NH_4^+ + OH^- = NH_3 + H_2O$	3×10^{10}
$NH_4^+ + aq = NH_3 + H^+$	$<6 \times 10^{-3}$
$CH_3COCH_3 + aq = CH_3COCH_2^- + H^+$	5×10^{10}
$Fe^{3+} + Cr^{2+} = Cr^{3+} + Fe^{2+}$	2×10^3
$Fe^{2+} + Co^{3+} = Fe^{3+} + Co^{2+}$	10
$I^- + OCl^- = OI^- + Cl^-$	2×10^8
$CH_3Br + I^- = CH_3I + Br^-$	10^{-4}
$CH_3I + OH^- = CH_3OH + I^-$	8×10^{-6}
$H^+ + CH_3COO^- = CH_3COOH$	$4{\cdot}5 \times 10^{10}$

There is an unexpected similarity between the values for k_2 in Tables 7.1 and 7.2. The maximum observed rate constant (k_2) in solution appears to be about 10^{10} mol^{-1} dm^3 s^{-1} which is not very different from the maximum rate constant in gases. It seems that the frequency of collision between reactant species *can* be nearly as high in solutions as in gases. This assumption is supported by the fact that some reactions occur at about the same rates in the gas phase as in solution. For instance, k_2 for the dimerization of cyclopentadiene at 323 K is $6{\cdot}5 \times 10^{-6}$ mol^{-1} dm^3 s^{-1} in the gaseous phase and $9{\cdot}0 \times 10^{-6}$ mol^{-1} dm^3 s^{-1} in a solution of tetrachloromethane.

One possible modification of the collision model which might explain these observations is as follows. The presence of the solvent molecules obstructs the collision of solute particles. However, once the two solute particles have met, the particles may remain close to one another for a considerable time, since they are surrounded by a 'cage' of solvent molecules which prevents their escape. Repeated collisions between the same pair of reactant particles may then occur. The net result may be that the effective collision frequency is not much different from that in the gas phase.

However, this modification of the model cannot be entirely satisfactory. There are some reactions, for instance, with rate constants which are greatly different in the gaseous phase and in solution, and many others which, although they cannot be investigated in the gas phase, proceed at very different rates in different solvents.

In order to derive an expression for the maximum predicted value of k_2 it is therefore necessary to take account of the properties of the solvent (viscosity, dielectric constant, etc.) which may affect the diffusion of solute particles. This can be done (see Appendix C) and the predicted value of k_2 is about 10^{10} mol^{-1} dm^3 s^{-1} which agrees well with the maximum observed values in Table 7.2. However, the derivation ignores, for instance, the possible effects of the molecular structure of the liquid and the specific solvation of the reactants, and can therefore only be an approximation, particularly for ionic reactions in aqueous solution. The complexity of solution reactions militates against such a rigorous treatment of reaction rates as that given to gaseous reactions.

The Arrhenius equation

In 1889, S. Arrhenius suggested the following relationship between rate constants and temperature.

$$k = A\exp(-E/RT) \qquad (5)$$

A and E are defined as temperature-independent constants,† T is the absolute temperature, and R is the Gas Constant. The fraction E/RT is defined as a pure number so that E has the same units as RT. These are J K^{-1} mol^{-1} $\times$ K, so that E has the energy units J mol^{-1}. The units of A are the same as those for the rate constant.

This equation was suggested by Arrhenius following work by van't Hoff and Guldberg and Waage. He introduced the additional idea that, for reaction to occur, the reactants must get into an 'activated state' in which they possess sufficient energy for reaction to occur. The rest of the molecules were considered to be non-activated and therefore incapable of reaction. Arrhenius was not able

† It is found that the values of A and E for a reaction are characteristic of that reaction.

to provide a detailed concept of the nature of the 'activated state' and was therefore unable to give precise theoretical meanings to the terms A and E. It is the attempts to provide these details which have led to all subsequent developments in the theory of reaction rates.

The justification of equation (5) is mainly experimental, not theoretical. Taking logarithms of equation (5),

$$\log_e k = \log_e A - \frac{E}{RT}$$

or

$$\log_{10} k = \log_{10} A - \frac{E}{2{\cdot}30RT} \qquad (6)$$

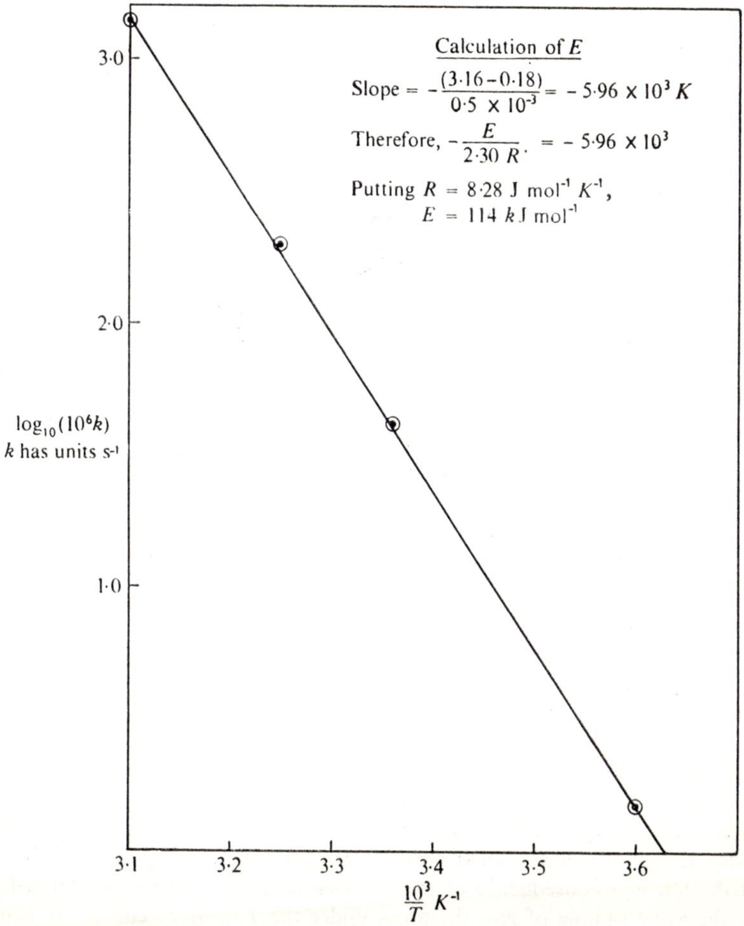

Fig. 7.1. An Arrhenius plot for the decomposition of benzene diazonium chloride. (from *J. Amer. Chem. Soc.* 1940, **62**, 1400)

An Arrhenius plot is a plot of $\log_e k$ or $\log_{10} k$ against $1/T$, which is linear if equation (5) applies. The slope of the straight line is $-E/R$ or $-E/2{\cdot}30R$ and the intercept on the y axis is $\log_e A$ or $\log_{10} A$. In Fig. 7.1 an Arrhenius plot is shown for the decomposition of benzene diazonium chloride. Such a good linear plot can be obtained for many, although not all, reactions.

Can the Arrhenius equation be interpreted using the collision model developed earlier in the chapter? In equation (5) A has the units of a rate constant and it is decreased by the factor $\exp(-E/RT)$ to give k, the experimental rate constant. It is therefore tempting to consider the exponential term as representing the fraction of collisions which are successful. Can this be justified in terms of the Kinetic Theory of Gases, that is, can an exponential term be introduced into equation (4)?

The exponential factor

It has already been suggested that a gas consists of molecules with widely different velocities and hence translational kinetic energies. By using probability theory, Maxwell (1860) and Boltzmann (1868) calculated expressions for the distribution

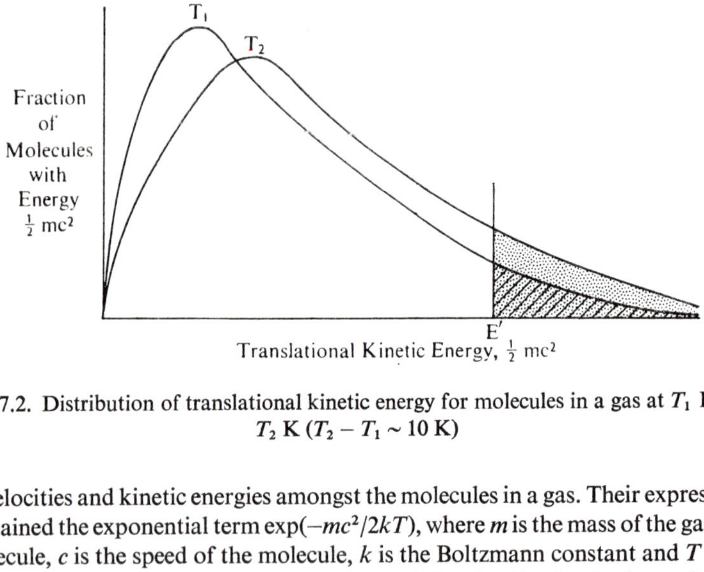

Fig. 7.2. Distribution of translational kinetic energy for molecules in a gas at T_1 K and T_2 K ($T_2 - T_1 \sim 10$ K)

of velocities and kinetic energies amongst the molecules in a gas. Their expressions contained the exponential term $\exp(-mc^2/2kT)$, where m is the mass of the gaseous molecule, c is the speed of the molecule, k is the Boltzmann constant and T is the absolute temperature of the gas. The form of the distribution is shown in Fig. 7.2 where the fraction of molecules with energy $\frac{1}{2}mc^2$ is plotted against $\frac{1}{2}mc^2$, the kinetic energy of translation of one molecule.

The graph shows that the majority of molecules have energies within a fairly limited range, but that a small proportion have higher energies. This kind of distribution is a consequence of the exponential term in the distribution law. For the same sample of gas, the areas under the T_1 and T_2 curves are equal. However, the area under the T_1 curve beyond E' is about half the corresponding area under the T_2 curve. Therefore a small rise in temperature (i.e. 10 K) has a very marked effect on the fraction of molecules with an energy greater than, or equal to, E' molecule^{-1} (or NE' mol^{-1}, where N is the Avogadro Number).

Now consider a gaseous mixture of A and B molecules which can react by collision of A and B molecules. When an A and a B molecule collide, the 'effective' collisional energy will depend on the direction of approach as well as on the individual kinetic energies. This effective energy will be at a maximum for a 'head-on' collision and will be much lower for glancing collisions. It can be shown that the fraction of collisions in which the effective kinetic energy exceeds a given value E_c mol^{-1} (or E_c/N molecule^{-1}) is $\exp(-E_c/RT)$ where R is the Gas Constant.

This result clearly shows the way in which an exponential term can be introduced into the predicted expression for the rate constant:

$$k_2 = YN \qquad \text{[Equation (4)]}$$

Here, the important assumption was that every collision between reactant molecules produces a reaction. This has now to be modified to: only those collisions between reactant molecules which involve an effective kinetic energy of E_c or greater will produce reaction. Equation (4) must be rewritten:

$$k_2 = YN\exp(-E_c/RT) \qquad (7)$$

This equation should be compared with the Arrhenius equation [Equation (5)]. Why a minimum energy is necessary for reaction may be explained by reference to the processes which are thought to occur when reactants combine and form products. This is discussed later in connection with the Transition State Theory.

There is now a theoretical basis for the observation that rate constants vary markedly with temperature: a 10 K rise in temperature, for instance, frequently increases the rate constant by between 50 per cent and 300 per cent (see Question 1 at the end of the chapter). This is explained as the variation with temperature of the fraction of collisions which involve an effective kinetic energy greater than or equal to E_c. Reference to Fig. 7.2 shows that the area under the curve beyond E_c is much larger for the curve at temperature T_2 than at T_1. It is therefore the shape of the distribution curve and its variation with temperature which accounts for the observation.

However, one serious limitation to the application of the theory is that it fails to provide a basis for the prediction of E_c for any given reaction.

The pre-exponential term in the Arrhenius equation

The term, A, in equation (5) can be determined experimentally and appears as the intercept, $\log_{10} A$, in a plot of $\log_{10} k$ against $1/T$. When $E = 0$ in the Arrhenius equation,

$$k = A\exp(-O/RT) = A$$

and A is now seen as the rate constant which would apply if all collisions were successful in promoting a reaction.

The application of the collision model has been shown to predict the value for k_2, the bimolecular second order rate constant, as YN [equation (4)], assuming that all collisions were successful. Hence it would be expected that for a bimolecular reaction,

$$A = YN \tag{8}$$

This can readily be tested by comparing the experimental value for A with the predicted value of YN. Notice particularly that the prediction suggests a value for A which should be much the same for all reactions. Table 7.3 shows how far this prediction agrees with experiment.

Table 7.3. Some A and E values for bimolecular gaseous reactions
$YN = A_{\text{predicted}} \sim 10^{11}\ \text{mol}^{-1}\ \text{dm}^3\ \text{s}^{-1}$

Reaction	E kJ mol^{-1}	$A_{\text{exp.}}$ mol^{-1} dm^3 s^{-1}	$A_{\text{exp}}/A_{\text{pred.}}$ (approx.)
$H + Br_2 = HBr + Br$	4·2	$1{\cdot}6 \times 10^{11}$	1
$H_2 + Br = HBr + H$	74	$4{\cdot}8 \times 10^{10}$	0·5
$HBr + Br = H + Br_2$	176	$3{\cdot}2 \times 10^{10}$	0·3
$2NOCl = 2NO + Cl_2$	99	$5{\cdot}0 \times 10^{9}$	0·05
$Cl + NOCl = NO + Cl_2$	4·2	$1{\cdot}1 \times 10^{10}$	0·1
$H_2 + Cl = HCl + H$	23	$7{\cdot}9 \times 10^{10}$	1
$CH_4 + Cl = CH_3Cl + H$	16	$2{\cdot}5 \times 10^{10}$	0·3
$CH_3 + CH_3 = C_2H_6$	0	$2{\cdot}0 \times 10^{10}$	0·2
$2CH_2{=}CH{-}CH{=}CH_2$ 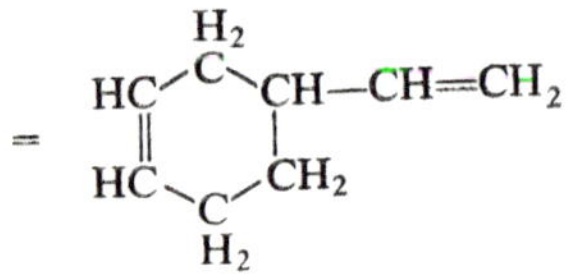	106	$4{\cdot}7 \times 10^{7}$	0·0005
$CH_2{=}CH{-}CH{=}CH_2 + CH_2{=}CH{-}CHO$ 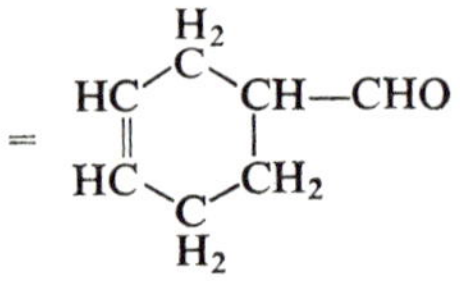	83	$1{\cdot}5 \times 10^{6}$	0·00001

Since YN was calculated using only approximate values of molecular weights and other properties of the molecules it should not be expected that very close agreement be obtained. Values between 10^{10} and 10^{11} mol^{-1} dm^3 s^{-1} can be considered to be in reasonable agreement, although the tendency for k_2 to be nearer 10^{10} than 10^{11} should be noted. The two most striking exceptions are the gaseous organic reactions where the experimental value for A is considerably less than 10^{10} mol^{-1} dm^3 s^{-1}. Both reactions differ from the others in the complexity of the reactant molecules. Similar discrepancies have been found for other reactions involving large molecules. The collision model needs modification to account for these large discrepancies and it has been suggested that an orientation (or steric) factor must be introduced.

Any collision needs a certain minimum energy, E_c, for reaction. In addition, the molecules may need to be correctly orientated with respect to each other so that the energy is applied in the right place. If they are not correctly orientated the collisional energy may not be sufficient for reaction to occur. Fig. 7.3 illustrates this by considering possible orientations for the collision of an atom A with a diatomic molecule AB. In (a) and (b), A and AB approach each other with an orientation suitable for forming A_2 and B. In (c), however, the orientation of the colliding molecules is such that there is likely to be no net change in the system.

For many reactions involving small molecules there is no need to introduce a correction factor (i.e. $A_{exp.}/A_{pred.} \sim 1{\cdot}0$) so that orientation effects in a reaction like that in Fig. 7.3 must be unimportant. However, for reactions involving

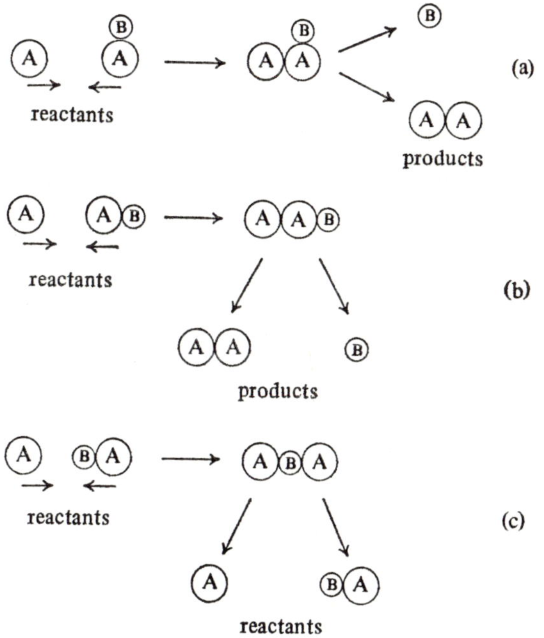

Fig. 7.3. Orientation effects on the reaction $A + AB = A_2 + B$

complex molecules $A_{exp.}/A_{pred.} \ll 1$ so that the orientation effect must be assumed to be very important. For large molecules there must be a relatively large number of orientations like that in Fig. 7.3 (c) where the products are unlikely to be formed, whereas there must be very few where conditions are favourable for the formation of a complex product. In other words, reaction often will not occur even when the collisional energy is greater than E_c because the molecules will be unsuitably orientated. The orientation factor is represented as p and equation (7) is rewritten as:

$$k_2 = pYN\exp(-E_c/RT) \tag{9}$$

As in the case of the value of E_c, there is unfortunately no way of predicting the magnitude of p for any reaction. A further difficulty with the above model for

p is that it cannot explain why p is sometimes greater than unity. The factor, p, therefore, is an arbitrary correction term for which there is only a partial qualitative explanation.

The usefulness of the collision theory

A summary can now be made of the similarities between the experimental Arrhenius equation (5) and the equation for the bimolecular rate constant, k_2, predicted from the collision theory [equation (9)].

$$k = A\exp(-E/RT) \quad (5)$$

$$k_2 = pYN\exp(-E_c/RT) \quad (9)$$

The value for YN can be calculated from theory, but only in some cases is it approximately equal to A. In some reactions an arbitrary factor, p, must be introduced so that the pre-exponential terms in the two equations can be equated. Values for E vary from reaction to reaction. Values for E_c cannot be predicted for any reaction so that E and E_c cannot be compared.

The term Y, as derived from the Kinetic Theory, is dependent on $T^{1/2}$ so that the pre-exponential term in equation (9) is temperature-dependent. However, experimental accuracy is usually not sufficient to distinguish this mild temperature dependence because of the very strong temperature dependence of the exponential term.

Diffusion-controlled reactions

For a gaseous reaction the maximum rate is determined by the maximum collision frequency (Z). The rate of reaction can only equal Z/N mol dm^{-3} s^{-1} when E (or E_c) is negligibly small.

i.e. rate of reaction $= k_2$[A][B]
(for a bimolecular reaction)

$$= pYN\exp(-E_c/RT)\text{[A][B]} \quad \text{[from Equation (9)]}$$

$$\sim pYN\text{[A][B]} \qquad \text{when } E_c \sim 0$$

For reactions in solution the motion of the reacting particles is affected by the medium (i.e. the solvent), and the rate of diffusion of reacting particles is slower than in gases. Therefore, the frequency of collisions in solution is limited by the rate at which the reactant particles can diffuse through the medium. When E (or E_c) is very small so that most collisions are successful, the rate of reaction is determined by the rate of diffusion. When E is large then the frequency of *successful* collisions becomes small and the rate of diffusion is no longer the factor which limits the rate of reaction. In practice it has been found that reactions which are *diffusion-controlled* have $E \leqslant 20$ kJ mol^{-1}.

The rate of diffusion provides the upper limit for reaction rates in solution. The maximum value for k_2 in solution reactions is about 10^{10} mol^{-1} dm^3 s^{-1} so that diffusion-controlled reactions are almost as fast as the fastest gaseous reactions. This value for k_2 is predicted approximately by considering the properties of the medium which control the rate of diffusion (see Appendix C). Some experimental values of k_2 are a little higher or lower than 10^{10} mol^{-1} dm^3 s^{-1} due to specific effects of the solute particles, particularly if they are ions. Some diffusion-controlled reactions can be identified in Table 7.2.

Bonding in chemical reactions

There are several defects in the simple collision theory outlined in this chapter. For example, molecules are considered to possess rotational and vibrational energy as well as translational kinetic energy, and yet the simple collision theory makes no allowance for these extra energy terms. Again, the assumption that all molecules can be considered as hard spheres is another obvious defect. Clearly, molecules have their own arrangement of atoms in space and their own particular distribution of electron density.

However, these defects must not be allowed to detract from the many merits of the collision theory. This chapter has shown how the collision theory has been very successful in providing a *picture* of how reactions occur, and partly successful in providing a *quantitative* basis. The assumptions inherent in the theory are obviously useful approximations.

The fact remains, however, that the collision theory offers no picture of the processes occurring *within* the molecules during the time when the reactant molecules are rearranging to product molecules. Bond making and bond breaking processes, which are so important to a modern view of chemical reactions, are ignored. The collisional energy, E_c, necessary for reaction is presumably dependent on such factors, and it clearly cannot be predicted using a theory based on the assumption of hard spheres. It is because of this fundamental gap in the collision theory that another mathematical formulation has been designed in an attempt to explain the experimental results.

The Transition State Theory

A detailed treatment of the Transition State Theory is beyond the scope of this book, but some of its qualitative aspects can be examined.

Consider the gaseous reaction A + BC = AB + C, where BC and AB represent diatomic molecules while A and C represent atoms. The sequence of events during this reaction can be postulated as:

Stage 1. A and BC approach one another.
Stage 2. When A and BC are close enough together, the electrons are rearranged leading to a weakening of the B—C bond and a partial formation of the A—B bond.
Stage 3. Either the products, AB and C, separate from one another or the reactants, A and BC, separate again. Thus, if reaction does occur:

A + B—C → A----B----C → A—B + C

(The full lines represent covalent bonds and the broken lines represent partially formed and partially broken covalent bonds.)

The Transition State Theory concentrates on the half-way stage (A----B----C) between reactants and products, which is termed the transition state or the *activated complex*. Further discussion is required to define this half-way stage satisfactorily.

Stage 1. As A and BC approach one another repulsive forces build up mainly between the electron clouds around each species. A certain amount of energy E_a, called the *activation energy*, is thus required to force A and BC to approach closely enough for the reaction to occur. This energy is provided by the kinetic energy of A and BC. As they approach, A and BC are slowed so that their kinetic energies decrease while the potential energy of the system increases. If the kinetic energies of A and BC (along the line of their approach) are smaller than E_a, the reactants will not be able to approach closely enough to attain the high potential energy arrangement in stage 2, and an elastic collision occurs. However, if the kinetic energy is equal to or greater than E_a, the arrangement represented in stage 2 can be attained.

Stage 2. Some or all of the kinetic energy which A and BC possessed before they approached one another has been converted into the potential energy of the activated complex. In the Transition State Theory, the activated complex is assumed to be unstable and to decompose spontaneously and without hindrance into either A and BC or AB and C. If the reactants are reformed, once again an elastic collision has occurred.

Stage 3. If AB and C are formed from the activated complex, the reaction has occurred and, as the products separate, potential energy is converted to kinetic energy. However, the products will not have the same total kinetic energy as the reactants. If the overall reaction is exothermic, AB and C will have more kinetic energy than A and BC, the change in kinetic energy arising from the difference between the potential energies of A and BC and AB and C.

The energy changes during the reaction can be shown diagrammatically (see Fig. 7·4). Because the Transition State Theory concentrates on the activated

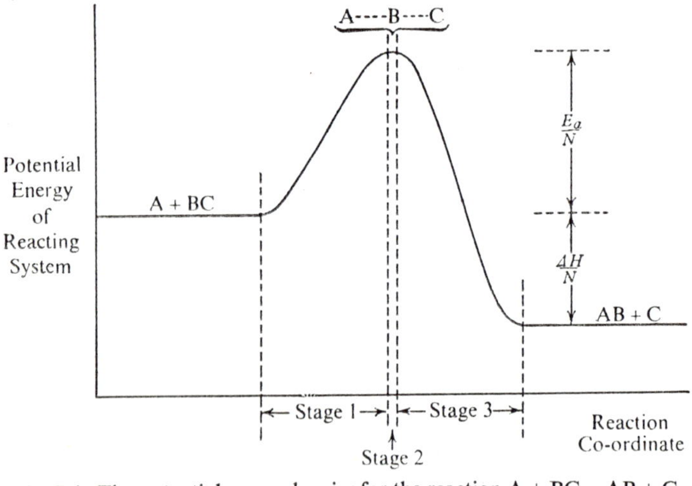

Fig. 7.4. The potential energy barrier for the reaction A + BC = AB + C

complex, it is concerned with the potential energy of the system and not with the kinetic energy from which it was derived. It is therefore usual to plot the potential energy of the system at the various stages in the reaction. The curve shown in Fig. 7.4 is for a head-on collision of A with BC, i.e. in the direction along the line through the nuclei. (A→←B—C)

The curve shown is for an exothermic reaction with a heat of reaction ΔH mol^{-1} (or $\Delta H/N$ molecule^{-1}). The maximum on the curve represents a potential energy barrier which must be crossed if the reaction is to occur. The activated complex can now be defined as the arrangement† of atoms and electrons at the peak of the potential energy curve. The activated complex has a very short life, because the slightest displacement of either A or C away from B causes it to decompose into either reactants or products respectively.

If the reaction is reversible, the same activated complex is involved, but for the reverse reaction, the potential energy barrier is higher $(E_a + \Delta H)$ mol^{-1} and the overall reaction is endothermic.

When the potential energy barrier is very low, the vast majority of collisions will involve molecules with sufficient kinetic energy for the activated complex to be formed. Under these conditions the rate is very nearly equal to the collision frequency of the reactant species. If such a reaction occurs in solution it is likely to be diffusion-controlled (see page 92).

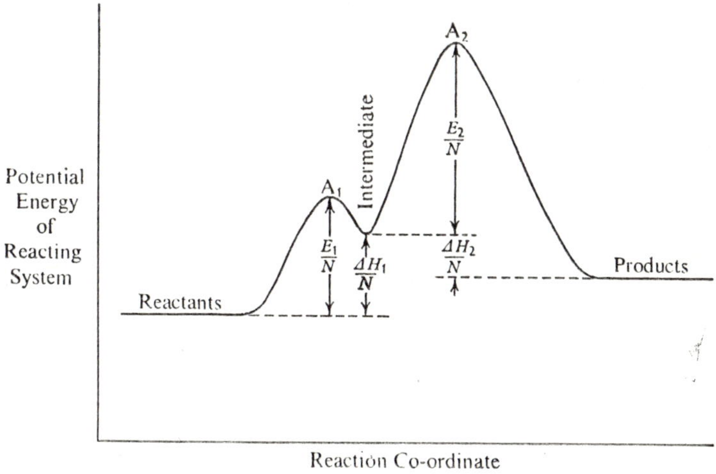

Fig. 7.5. A possible potential energy curve for a two-step reaction

A complex reaction proceeds by the formation and decomposition of one or more intermediate compounds. In this case there is an activated complex for each step of the reaction. A two-step reaction could have a potential energy curve as in Fig. 7.5. The reactants form the intermediate compound through an

† Hence Transition State Theory takes entropy into account.

endothermic reaction (ΔH_1 mol^{-1}) with an activation energy E_1 mol^{-1} and activated complex A_1. The intermediate is a normal molecule but is thermodynamically unstable because of the low activation energy (($E_1 - \Delta H_1$) mol^{-1}) necessary for its decomposition to the reactants. It can be converted into reactants again or it can, through another type of collision, react in the second step to form the products through the activated complex A_2. This step may or may not involve another reactant molecule. The overall reaction is endothermic (($\Delta H_1 - \Delta H_2$) mol^{-1}).

The potential energy curve shows that the second step may be the slow one, because it has the larger activation energy. Thus kinetic experiments on this reaction would give a rate expression for the second step.† In this way, specific information about the composition of the activated complex for the slow step is obtained. For example, the rate expression for the reaction of bromide and bromate ions in acid solution is:

$$-\frac{d[BrO_3^-]}{dt} = k_4[H^+]^2[Br^-][BrO_3^-]$$

which shows that the activated complex for the slow step contains two hydrogen ions, one bromide and one bromate ion. Hence the species which react in this step contain between them these four ions. A reasonable guess is that there are two rapid and reversible previous steps (with low activation energies) involving the formation of the ion-pairs H^+Br^- and $H^+BrO_3^-$ (see page 77).

The mathematical treatment of the Transition State Theory provides an expression for the rate constant of an elementary reaction in terms of the properties of the activated complex. This expression for k_2 has a form similar to the experimental Arrhenius equation and to the theoretical equation derived from the collision theory (see Appendix C). Although the mathematics are formidable, an estimate of the activation energy has been made for one or two reactions involving simple species. (Collision theory is incapable of this). The calculation can be done by solving the Schrödinger wave equation for the particular reacting system. For more complex reacting species the calculations are prohibitively long and very difficult. Several approximations can be made but they all use arbitrary adjustments and the predicted rates of reaction are often rather different from the experimentally observed rates. Therefore, the activation energy E_a is *defined* as the minimum energy required for an elementary step to occur, but the theory can rarely make any *quantitative* predictions of E_a. This emphasizes that at present, the main use of the Transition State Theory is not as an exact procedure for calculating rates, but rather as a means of picturing the activated complex in terms of modern ideas on bonding and structure.

Because there is no way of predicting E_a for most reactions, it is not possible to compare E_a with the energy term in the Arrhenius equation. Even if such a comparison were possible there are reasons why the two values may not agree. For instance, many reactions are complex and yet a single energy term (E) is calculated from the experimental results. It is doubtful if a precise meaning can be given to such a term for a complex reaction. It is now well established that the Arrhenius equation does not hold for some complex reactions. It can be said

† There are possible exceptions to this: the pre-exponential factors also contribute to the relative rates.

that when a plot of $\log_{10} k$ against $1/T$ is not linear, this indicates that the reaction occurs in several steps. However, the converse of this is not necessarily true because many complex reactions do give a linear plot.

Questions

1. It is often said that, for many reactions, the rate approximately doubles for an increase in temperature of 10 K around room temperature. Show that this only applies if the Arrhenius energy (E) is about 50 kJ mol^{-1} ($R \sim 8$ kJ mol^{-1} K^{-1}).
2. Devise ways of obtaining the Arrhenius energies (E) of some of the reactions described in Chapter 4. In view of the fact that these reactions are probably all complex, discuss what significance the values for E may have.
3.(*a*) Slow reactions in solution are unaffected by a change in the viscosity of the medium whereas very fast reactions in solution are slowed by an increase in viscosity.
(*b*) Slow reactions in solution are markedly affected by an increase in temperature whereas very fast reactions are only slightly affected.
Explain these two observations using the concept of diffusion-controlled reactions.
4. Use the results in Experiment 4.1 to plot $\log_{10} k$ against $1/T$ for the decomposition of benzene diazonium chloride in aqueous solution. Hence determine E and A in $k = A\exp(-E/RT)$. Estimate the accuracy of each determination, and compare the value of E with that obtained in Fig. 7.1.

8 More about catalysis

A catalyst may be defined as a substance which increases the rate of a reaction and is recoverable unchanged in amount at the end of the reaction. This definition enables a catalyst to be recognized experimentally. If it becomes possible to propose a theory of how catalysts work it may then be possible to define catalysts in theoretical terms as well as experimentally.

Some familar examples of catalysis, most of which have been mentioned earlier in this book, are given in Table 8.1. In each case the rate of reaction is increased when the catalyst is added, and usually the increase depends on the amount or concentration of the catalyst. For example, the iodination of acetone is first order with respect to hydrogen ions but hydrogen ions do not appear as a reactant in the stoichiometric equation because they are recoverable unchanged at the end of the reaction.

A. Intermediate compounds in catalysed reactions

In Chapter 6 it has been explained that most reactions are thought to occur in a series of elementary steps. In order to explain catalysis it could be postulated that catalysts enter into the reaction they are catalysing. For this idea it is necessary to postulate one step in which the catalyst reacts and another in which it is regenerated. No catalysed reaction could therefore consist of a single step but should contain at least two steps.

For the reaction whose stoichiometric equation is A + B = products which is catalysed by substance C, a simple mechanism could be:

A + C = AC			Step 1
reactant	catalyst	intermediate	
AC + B = products + C			Step 2
	reactant		

The addition of these two steps produces the stoichiometric equation for the reaction, in which the catalyst does not appear.

If step 1 is rate determining, the predicted rate expression would be

$$-\frac{d[B]}{dt} = -\frac{d[A]}{dt} = k'[A][C] \qquad (1)$$

and the reaction would be first order in the catalyst, but zero order in B, one of the reactants.

If step 1 is not very much slower than step 2, the situation is more complicated and is not dealt with here. However, if step 2 is a much slower step than step 1, a simplified treatment is possible and is given below. (For a fuller account the reader should consult Appendix F.)

When step 1 is reversible and fast in both directions, and step 2 is slow and

Table 8.1. Some familiar examples of catalysis

Reaction	*Catalyst*	*Comments*
$2KClO_{3(c)} = 2KCl_{(c)} + 3O_{2(g)}$	$MnO_{2(c)}$	Other oxides also catalyse this decomposition e.g. CuO, ZnO, NiO, Fe_2O_3
$CH_3COCH_{3(aq)} + I_{2(aq)} = CH_3COCH_2I_{(aq)} + H^+_{(aq)} + I^-_{(aq)}$	$H^+_{(aq)}$	Rate expression is: $-\frac{d[\text{acetone}]}{dt} = k_2[\text{acetone}][H^+]$ Potentially autocatalytic (H^+_{aq})
$2H_2O_{2(aq)} = 2H_2O_{(l)} + O_{2(g)}$	$MnO_{2(c)}$ / colloidal $Fe(OH)_3$ / catalase (and other enzymes)	—
$4NH_{3(g)} + 5O_{2(g)} = 4NO_{(g)} + 6H_2O_{(l)}$	Cu / Pt	Important in manufacture of nitric acid
$2H_{2(g)} + O_{2(g)} = 2H_2O_{(l)}$	Pt	—
$2MnO^-_{4(aq)} + 16H^+_{(aq)} + 5C_2O^{2-}_{4(aq)} = 2Mn^{2+}_{(aq)} + 10CO_{2(g)} + 8H_2O_{(l)}$	$Mn^{2+}_{(aq)}$	Autocatalytic

irreversible, it can be assumed that step 1 achieves equilibrium despite the slow removal of AC through step 2.

$$\mathrm{A + C \rightleftharpoons AC} \qquad \text{Step 1 (fast in both directions)}$$

$$\mathrm{AC + B \xrightarrow{k_2} products + C} \qquad \text{Step 2 (slow)}$$

An equilibrium concentration of the intermediate, AC, is maintained throughout the reaction, and the equilibrium constant, K, for step 1 is:

$$K = \frac{[\mathrm{AC}]}{[\mathrm{A}][\mathrm{C}]} \qquad (2)$$

The rate of reaction is the rate at which the products form, and is given by:

$$+\frac{d[\text{products}]}{dt} = k_2[\mathrm{AC}][\mathrm{B}] \qquad (3)$$

Substituting for [AC] in (3) gives:

$$+\frac{d[\text{products}]}{dt} = k_2 K[\mathrm{A}][\mathrm{B}][\mathrm{C}] \qquad (4)$$

and the reaction is first order in all three species.

A rate expression like equation (1) is obtained for the iodination of acetone in acid solution (see Table 8.1). Equations (1) and (4) explain how the concentration of the catalyst may appear in the rate expression. Much more complicated examples are possible.

Intermediates, like AC, formed by a reactant molecule and a catalyst molecule clearly cannot be produced during the corresponding uncatalysed reaction. The mechanisms of the catalysed and uncatalysed reaction must be different. For some reactions, the uncatalysed change is a single elementary step whilst the catalysed change is a complex process; for others, both changes are complex processes.

The identification of intermediates is as important in postulating likely mechanisms for catalysed reactions as it is for uncatalysed reactions. Experiments 8.1 and 8.2 illustrate the existence of clearly visible intermediates, and some guesses about possible mechanisms can be made.

Experiment 8.1 The 'blue bottle' experiment†

Reagents: glucose
sodium hydroxide pellets
1 per cent alcoholic methylene blue solution.

Procedure

Dissolve about 10 g of glucose and 10 g of sodium hydroxide in about 300 cm^3 of water in a 500 cm^3 flask fitted with a rubber bung. Add 1 cm^3 of methylene blue solution, replace the bung and allow the mixture to stand until the colour disappears. Shake the flask vigorously for a few seconds and then allow it to stand. Repeat the procedure at least once more.

Devise experiments which will help to determine the conditions under which the appearance and disappearance of the blue colour

† See *J. Chem. Ed.*, 1963, **40**, 578.

occur. Some suggestions are listed below. For each of these tests it is better to withdraw a small sample of the reaction mixture and perform the experiment in a boiling tube fitted with a rubber bung.

(*a*) Is the blue colour due to a reaction between the mixture and the glass of the flask or the bung?

(*b*) Does warming produce the blue colour? On the basis of this answer is it likely that the increase in kinetic energy of the mixture on shaking the flask produces the colour?

(*c*) Is the blueing and deblueing cycle caused by the reaction being reversible?

(*d*) Does air have any effect? After thoroughly shaking the flask, gently lift the bung from the neck of the flask. When the blue colour has disappeared observe carefully the meniscus of the mixture. Bubble oxygen rapidly into the mixture, and then repeat using other gases.

(*e*) Does the intensity of the blue colour depend on the length of time the flask is shaken?

(*f*) Investigate the times taken for the blue colour to appear and to disappear. Is either of these dependent on the length of time the flask is shaken?

(*g*) What happens when more water is added? By how much is the rate of deblueing affected by diluting the mixture with an equal volume of water?

Discussion of results

Much of the experimental evidence indicates that oxygen from the air is involved in producing the blue colour. It seems, therefore, that the blue compound is the oxidized form of the dye, methylene blue. This compound slowly reacts to produce a colourless solution. Presumably one of the other reactants, or a derivative of them, is able to reduce the blue dye. There are at least two observations which suggest that this blueing–deblueing cycle is unlikely to be one reversible reaction. What are they?

A possible mechanism for this reaction is therefore:

$O_{2(g)} \rightarrow O_{2(aq)}$ Step 1

$O_{2(aq)} + \underset{\text{methylene blue}}{M} \rightarrow \underset{\text{blue compound}}{B}$ Step 2

$B + \underset{\text{glucose}}{G} \text{ (or } OH^-) \rightarrow \text{products}$ Step 3

Since the blue colour appears rapidly, steps 1 and 2 must be relatively fast. The fading of the blue colour takes considerably longer suggesting that step 3 is the slow step.

This mechanism does not explain all the observations. Although only a very small amount of methylene blue is added the blueing–deblueing cycle can be repeated many times without the intensity of the blue colour fading. This suggests that M may be regenerated in step 3, i.e.

$$B + G(\text{or } OH^-) \rightarrow M + \text{products.}$$

This postulate is confirmed by two further observations. The intensity of the blue colour is independent of the length of time the flask is shaken unless that time is very short, and the time for deblueing is approximately proportional to the time of shaking. The more the flask is shaken the more oxygen gas is dissolved in the solution. A larger concentration of dissolved oxygen might be expected to increase the reaction rate. However, the rate of deblueing is *decreased.* This difficulty can be resolved by suggesting that the methylene blue is a limiting reagent. After sufficient oxygen has been dissolved to react with this limited amount of M no more blue colour (B) can be produced until step 3 has slowly removed some of B. If the concentration of B is to remain constant whilst B is reacting in step 3, however, it follows that M must be regenerated to provide reactant for step 2.

M can now be seen as the catalyst for the reaction since it is not consumed and therefore does not appear in the stoichiometric equation. The blue compound, B, is an intermediate formed by M and the dissolved oxygen, one of the reactants.

Dilution of the contents of the flask in Experiment 8.1(*g*) should show what effect concentration changes have on the total order of the reaction. The kinetics can be more thoroughly investigated (see Exercise 5).

Experiment 8.2 Oxidation of a tartrate by hydrogen peroxide (cobalt(II) ions as catalyst)

Procedure

Dissolve about 2 g of sodium potassium tartrate in about 50 cm^3 of water in a beaker. Warm to about 80°C and then add 10 cm^3 of '20 vol' hydrogen peroxide. Add a few crystals of cobalt(II) chloride and stir to dissolve.

Note the colour changes, and when no further change occurs add a few more cm^3 of hydrogen peroxide. Is it possible to isolate the intermediate compound so that it does not rapidly decompose?

Repeat the procedure using 10 cm^3 of water instead of hydrogen peroxide, and then using 10 cm^3 of peroxide but no tartrate. Investigate whether similar effects can be achieved using tartaric acid instead of the tartrate, and then cobalt(II) nitrate instead of cobalt(II) chloride.

Discussion of results

The pink cobalt(II) ion acts as a catalyst. The evidence suggests that the intermediate compound is formed from a cobalt(II) ion and *both* a hydrogen peroxide molecule and a tartrate ion. A possible mechanism for the reaction might be:

fast: H_2O_2 + tartrate anion $\rightarrow$ intermediate 1

fast: intermediate 1 + Co^{2+} $\rightarrow$ intermediate 2

slow: intermediate 2 $\rightarrow$ $CO_2 + O_2 + Co^{2+}$
(gases evolved)

The composition of the second intermediate, which must contain cobalt, is not known.

Intermediates are not usually so readily observed as in the above experiments. More refined techniques involving gas chromatography, radio-isotopes, or spectroscopy have to be used.

The Friedel–Crafts reaction is an example.

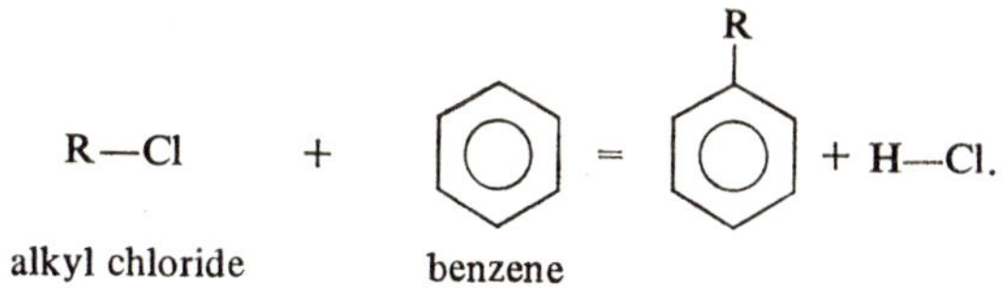

This reaction is catalysed by anhydrous aluminium chloride. The existence of an intermediate like $R^+(AlCl_4)^-$ is indicated by the fact that isotopically labelled chlorine atoms in $AlCl_3$ are exchanged with chlorine atoms in RCl. A possible mechanism is:

$$R{-}Cl + AlCl_3 \rightleftharpoons R^+(AlCl_4)^-$$ (this step accounts for the exchange)

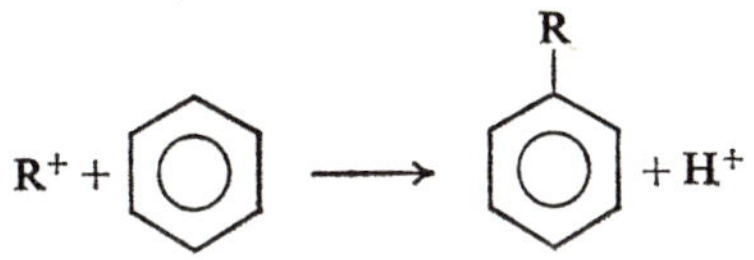

$$AlCl_4^- + H^+ \longrightarrow AlCl_3 + HCl.$$

B. Activation energy, E_a, in catalysed reactions

Why do catalysts increase the rate of a reaction so markedly?

It was established in Chapter 7 that any reaction rate is partly controlled by the height of the potential energy barrier, E_a; the lower the barrier, the more energetically favourable the pathway. A suggestion can be made, therefore, that a catalysed reaction has a lower E_a than the uncatalysed reaction. This can be tested by deter-

Table 8.2. *E* for catalysed and uncatalysed reactions. (Some values are only approximate)

	E kJ mol^{-1}		
Reaction	*Uncatalysed*	*Catalysed*	*Catalyst*
$2HI_{(g)} = H_{2(g)} + I_{2(g)}$	180	100	Au
		59	Pt
$2N_2O_{(g)} = 2N_{2(g)} + O_{2(g)}$	240	120	Au
		130	Pt
$CO(NH_2)_{2(aq)} + 2H_2O_{(l)} + H^+_{(aq)} = 2NH^+_{4(aq)} + HCO^-_{3(aq)}$	140	38	urease (an enzyme)
$2H_2O_{2(aq)} = 2H_2O_{(l)} + O_{2(g)}$	75	50	colloidal Pt
		21	catalase (an enzyme)
$2C_2H_5OC_2H_{5(g)} = 2CO_{(g)} + 4CH_{4(g)} + C_2H_{4(g)}$	220	140	$I_{2(g)}$
$2NH_{3(g)} = N_{2(g)} + 3H_{2(g)}$	380	160	W

mining values of the experimental activation energy, *E*, for the catalysed and uncatalysed reactions using Arrhenius plots (see page 87). It is assumed that such values will only be an *indication* of the heights of the barrier (E_a) for reasons already discussed in Chapter 7. Some results are given in Table 8.2.

The values of *E* for the catalysed reactions are indeed less than those for the uncatalysed reactions suggesting that the heights of the barrier (the activation energy, E_a) are lower too. In many catalysed reactions the concentration of the catalyst is rather low compared with the reactant concentrations, suggesting that reactant–reactant collisions are far more common than reactant–catalyst collisions. The catalysed pathway can then be favoured only if there are more *successful*

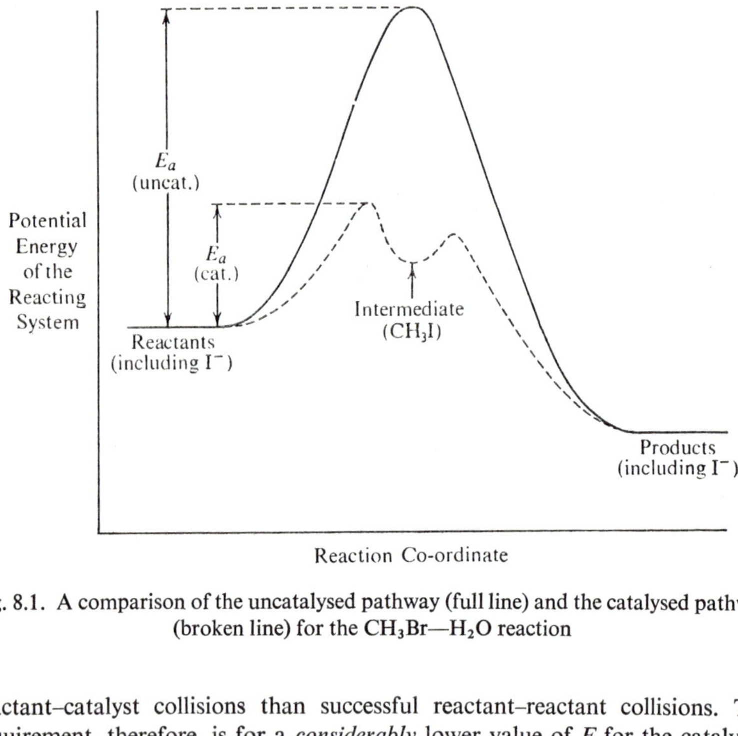

Fig. 8.1. A comparison of the uncatalysed pathway (full line) and the catalysed pathway (broken line) for the $CH_3Br—H_2O$ reaction

reactant–catalyst collisions than successful reactant–reactant collisions. The requirement, therefore, is for a *considerably* lower value of *E* for the catalysed reactions. The data in Table 8.2 do indicate that the activation energy for the catalysed reaction is usually very much lower than that of the uncatalysed reaction.† The enzyme-catalysed reactions, in particular, have very low values for *E*. Of course, if the catalyst concentration is comparable with the reactant concentrations there is no need for such a large reduction in *E*.

Fig. 8.1 shows the potential energy curves for a particular reaction, the hydrolysis of bromomethane by water, which can be catalysed by iodide ions.

$$H_2O_{(l)} + CH_3Br_{(aq)} = CH_3OH_{(aq)} + H^+_{(aq)} + Br^-_{(aq)}$$

† A full discussion would also take account of the pre-exponential terms.

When uncatalysed, the reaction is thought to proceed by one step involving collisions between water and bromomethane molecules. The activation energy, E_a, for this reaction is high and the rate of reaction is therefore low. The catalyst introduces a new pathway involving CH_3I as an intermediate.

$$I^- + CH_3Br \rightarrow CH_3I + Br^- \qquad \text{Step 1}$$

$$H_2O + CH_3I \rightarrow CH_3OH + H^+ + I^- \qquad \text{Step 2}$$

The activation energy, E_a, of each of these steps is very low and both proceed more rapidly than the reaction between bromomethane and water. The first step has the slightly higher E_a so that this may be rate determining. The 'valley' between the two 'hills' (the activated complexes) represents the potential energy of the intermediates in the catalysed reaction.

The introduction of the catalyst creates a second route. The reaction could still proceed via the uncatalysed pathway or it could proceed via the new pathway. The route involving the lower energy barrier is always favoured, even though there are two barriers to surmount instead of one, because there are more collisions with the energy required for the reaction to occur by this route.

An analogy could be drawn with the choice facing the owner of an old, unreliable car wishing to visit a nearby town. Two routes are open to him. One involves a tortuous ascent up a steep mountain pass, while the other runs over undulating but fairly low-lying land. The car is more likely to arrive at the destination if it is not over-stretched on the steep ascent. The same consideration applies on the return journey.

This last point helps to illustrate the observation that, for reversible reactions, catalysts for the forward reaction are also catalysts for the reverse reaction. In particular, it is found that the value of the equilibrium constant for a reversible reaction is not affected by the presence of a catalyst. Therefore, at equilibrium the catalyst must increase the rates of the forward and reverse reactions by the same factor. (If this did not occur, the Law of Conservation of Energy would be violated and a perpetual motion machine would be possible.)

C. Some types of catalysis—the specificity of catalysts

Traditionally, catalysed reactions are divided into the *homogeneous* type where the catalyst is in the same phase as the reactants, and the *heterogeneous* type where the catalyst is in another phase. Along with this separation has gone a different approach in the theoretical background to each type. However, this is an artificial separation because all catalysts appear to function by interacting with the reactants and usually forming definite chemical compounds (i.e. intermediates), thereby providing a new pathway for the reaction. Moreover, the distinction is also blurred by enzyme catalysis which is difficult to place in either category because enzymes are present as hydrosols.

However, there are some differences between the various types, and the important features of these are discussed below. Many of them illustrate that catalysts are often very specific in their action.

Catalysed reactions in solution (mainly homogeneous)

Redox reactions

In these reactions a suitable catalyst is an ion or molecule which can be oxidized by one of the reactants and subsequently regenerated by the reducing action of the other reactant (see Experiment 8.3 below).

In a few cases, one reactant is able to perform both the oxidation and reduction alternately. Hydrogen peroxide acts like this when it is catalytically decomposed to water and oxygen. Suitable catalysts are iron(III) ions, iodide ions, bromine molecules in acid solution, and solid manganese(IV) oxide. For instance, the two likely steps in the mechanism when bromine in aqueous acidic solution acts as the catalyst are:

$Br_2 + H_2O_2 \rightarrow 2Br^- + 2H^+ + O_2$
(hydrogen peroxide is oxidized by bromine)

$2Br^- + H_2O_2 + 2H^+ \rightarrow Br_2 + 2H_2O$
(hydrogen peroxide is reduced by bromide ions and hydrogen ions)

In the case of the manganese(IV) oxide, the reactant is in solution whilst the catalyst is a solid (i.e. heterogeneous catalysis). However, in this case it may be supposed that the hydrogen peroxide oxidizes manganese(IV) to a higher valency state, then reduces it back to manganese(IV). This reaction can only occur at the surface of the solid catalyst and not, as in the homogeneous catalysis, throughout the reaction mixture.

Experiment 8.3 Catalysis of the reaction between iron(III) ions and iodide ions in acid solution

$$2Fe^{3+}_{(aq)} + 2I^-_{(aq)} = 2Fe^{2+}_{(aq)} + I_{2(aq)}$$

Procedure

Prepare two solutions in separate 250 cm³ flasks using these quantities in each flask:

50 cm³ pure water
about 0·5 g potassium iodide
25 cm³ 2M hydrochloric acid
2 cm³ starch.

To *one* flask, add 2 drops of bench copper sulphate solution. Copper(I) ions will be produced by reaction with iodide ions.

$$2Cu^{2+}_{(aq)} + 2I^-_{(aq)} = 2Cu^+_{(aq)} + I_{2(aq)}$$

Pipette 5·0 cm³ of 0·1 M sodium thiosulphate solution into each flask, and then add 5·0 cm³ of bench iron(III) chloride solution from small measuring cylinders to each flask simultaneously. Shake thoroughly and place each flask on some white paper. Note which solution goes blue first.

Discussion of results

This can be interpreted as catalysis by copper(I) ions. The catalyst is oxidized to copper(II) ions by iron(III) ions and then reduced to copper(I) ions by iodide ions. What is the purpose of the sodium thiosulphate?

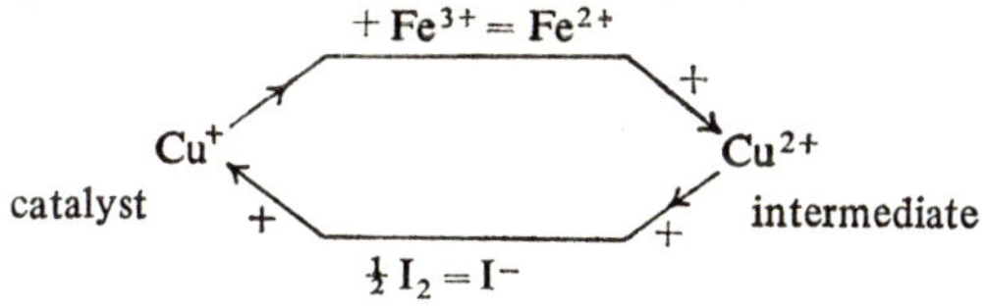

Acid-base catalysis

This may be the most common type of catalysis in aqueous solution since, on the Brönsted–Lowry acid-base theory, many molecules and ions as well as hydrogen ions and hydroxide ions are considered to be either acids or bases. Water itself is weakly amphoteric.

A large number of reactions, particularly in organic chemistry, are known to be subject to catalysis by either acidic or basic species or by both. If the catalysis can be shown to occur only with hydrogen ions, it is known as *specific catalysis by hydrogen ions*, whereas, if any acid (within the Brönsted–Lowry definition) catalyses the reaction, it is termed *general acid catalysis*, and so on. Table 8.3 gives some examples of specific and general catalysis.

In catalysis by an acid the first step is frequently thought to be the protonation (the addition of H^+) to an atom with a high electronegativity. This is probably the first step in the mechanism for the iodination of acetone in aqueous acidic solution.

$$CH_3\text{—}\underset{\underset{O}{\|}}{C}\text{—}CH_3 + H^+ \rightarrow CH_3\text{—}\underset{\underset{+OH}{\|}}{C}\text{—}CH_3$$

or

$$CH_3\text{—}\underset{\underset{O}{\|}}{C}\text{—}CH_3 + CH_3COOH \rightarrow CH_3\text{—}\underset{\underset{+OH}{\|}}{C}\text{—}CH_3 + CH_3COO^-$$

In base catalysis the first step is usually the removal of an 'acidic' hydrogen atom, as for example in the base catalysed iodination of acetone.

$$CH_3\text{—}\underset{\underset{O}{\|}}{C}\text{—}\underset{\underset{\text{(H)}}{|}}{CH_2} + OH^- \longrightarrow CH_3\text{—}\underset{\underset{O}{\|}}{C}\text{—}CH_2^- + H_2O$$

slightly acidic

or

$$CH_3\text{—}\underset{\underset{O}{\|}}{C}\text{—}CH_3 + CH_3COO^- \rightarrow CH_3\text{—}\underset{\underset{O}{\|}}{C}\text{—}CH_2^- + CH_3COOH$$

Surface catalysis (heterogeneous)

This includes mainly gaseous reactions catalysed at solid surfaces, but there are a few reactions in solution catalysed in this way, e.g. the use of manganese dioxide

Table 8.3

Reaction	*General or Specific*	*Catalytic Species* — *Basic*	*Catalytic Species* — *Acidic*
$NH_2NO_{2(aq)} = N_2O_{(g)} + H_2O_{(l)}$ (nitramide)	General base catalysis	Any base e.g. $OH^-_{(aq)}$, $CH_3COO^-_{(aq)}$	None
$CH_3COCH_{3(aq)} + I_{2(aq)} = CH_3COCH_2I_{(aq)} + H^+_{(aq)} + I^-_{(aq)}$	General acid–base catalysis	Any base	Any acid e.g. $H^+_{(aq)}$, $CH_3COOH_{(aq)}$
β-D-glucose = α-D-glucose (mutarotation of glucose)	General acid–base catalysis	Any base	Any acid
Inversion of sucrose (hydrolysis in acid solution)	Specific catalysis by $H^+_{(aq)}$	None	$H^+_{(aq)}$ only
Acid or alkaline hydrolysis of esters	Specific catalysis by $H^+_{(aq)}$ and $OH^-_{(aq)}$	$OH^-_{(aq)}$ only	$H^+_{(aq)}$ only
Depolymerization of diacetone alcohol	Specific catalysis by $OH^-_{(aq)}$	$OH^-_{(aq)}$ only	None

in the decomposition of hydrogen peroxide. Many of these reactions are important industrially (see Table 8.4.)

Table 8.4. Some industrial processes using surface catalysts.

Process	*Catalyst(s)*
$N_2 + 3H_2 = 2NH_3$ (Haber Process)	Fe. Efficiency increased by use of K_2O or Al_2O_3 as promoters
$CO + 2H_2 = CH_3OH$	ZnO and Cr_2O_3
$2SO_2 + O_2 = 2SO_3$ (Contact Process)	V_2O_5. Pt can be used but is very easily poisoned
Hydrogenation of >C=C< as in manufacture of margarine from oils	Finely divided Ni
$4NH_3 + 5O_2 = 4NO + 6H_2O$	Pt
Cracking mineral oil to petroleum spirit	A natural or synthetic alumino-silicate catalyst used in a fluidized bed

It is sometimes found that one solid or group of solids catalyses a certain type of reaction. For example, ethanol vapour can decompose in a number of ways.

Experiment 8.4 Catalytic decomposition of ethanol

(*a*) *Dehydration*

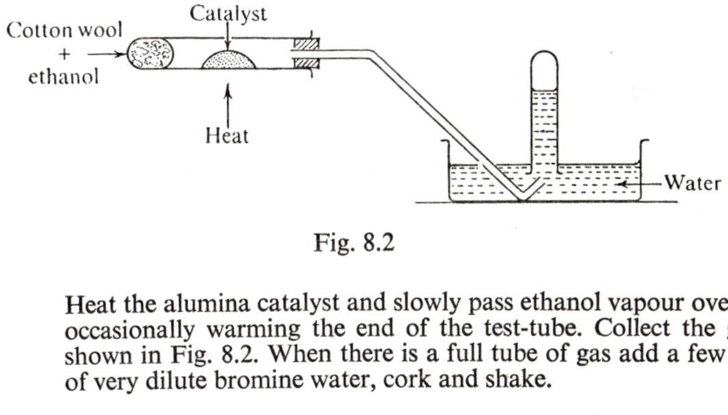

Fig. 8.2

Heat the alumina catalyst and slowly pass ethanol vapour over it by occasionally warming the end of the test-tube. Collect the gas as shown in Fig. 8.2. When there is a full tube of gas add a few drops of very dilute bromine water, cork and shake.

(*b*) *Dehydrogenation*

Use the same method as (*a*) but collect the product(s) by passing them down into a test-tube surrounded by a freezing mixture. Also use freshly reduced copper powder (or wire) as the catalyst instead of alumina. After a few cm^3 of liquid have been collected divide into two equal portions. Test one portion with Schiff's reagent and the other with Fehling's solution.

Is it possible to detect any acetaldehyde in (*a*) or any ethylene in (*b*)?

Discussion of results

Ethanol vapour can decompose in three main ways at temperatures around 300°C using a catalyst of either copper or alumina.

$C_2H_5OH_{(g)} = C_2H_{4(g)} + H_2O_{(g)}$	Al_2O_3 catalyst	Above 300°C
$C_2H_5OH_{(g)} = CH_3CHO_{(g)} + H_{2(g)}$	Cu catalyst	Between 200–300°C
$2C_2H_5OH_{(g)} = C_2H_5\text{—}O\text{—}C_2H_{5(g)} + H_2O_{(g)}$	Al_2O_3 catalyst	Between 260–300°C

The uncatalysed decomposition of ethanol only occurs at about 600°C or above, when the main product is acetaldehyde with some carbon monoxide and methane.

Such a diversity of products, depending on the temperature and presence of a catalyst, seems difficult to interpret. However, one way of accounting for the results is to accept that the molecules of a gas such as ethanol, at an elevated temperature, may decompose to give a number of products. In other words a number of reactions can occur at the same time. Although stoichiometric equations can be written for the individual reactions, it is frequently difficult in such cases to write an equation for the overall change which occurs. The catalyst serves to help one particular reaction out of the many possibilities. The different catalysts favour different products because they interact differently with the ethanol molecules. Some examples of the specificity of surface catalysts are given in Table 8.5.

Table 8.5. The specificity of surface catalysts in gaseous reactions

Reactions	*Catalysts*
Hydrogenation, Dehydrogenation	Transition metals like Fe, Ni, Pt, Pd, Cu, and Ag
Dehydration	Refractory oxides like Al_2O_3, SiO_2, MgO
Oxidation, Reduction	Non-refractory oxides like ZnO, CuO, Cr_2O_3, NiO

Adsorption

The interaction between solids and gases such as those in Table 8.5 is called adsorption. Two types of adsorption can be distinguished from measurements of the heat changes which accompany the process. The heat change for the first type is approximately equal to the latent heat of condensation of the gas (up to about 20 kJ mol^{-1}). In the second type the heat change is larger (about 100 to 400 kJ mol^{-1}) and is of the order of the heat change for a chemical reaction.

The inferences from these observations are that, in the first type, the change is essentially a physical one, whereas the second type involves the making and breaking of chemical bonds. The two types of adsorption are termed physical adsorption and chemisorption respectively.

A qualitative idea of how chemisorption occurs may be obtained by considering the bonding in a solid. In the body of the solid, each atom is surrounded by other atoms and the valence electrons of each atom are used in the bonding between the atoms. At the surface, however, the atoms are only partly surrounded by other atoms and have their valency requirements unfulfilled, i.e. they possess one or more free valence electrons. If a gaseous molecule approaches the surface there is the possibility of a compound being formed between the molecule and one or several surface atoms. A monomolecular layer of the reactant hence forms on the surface of the catalyst.

The mechanism of surface catalysis

Transition metals are usually good catalysts for reactions involving hydrogen, and hydrogen undergoes chemisorption on most of them. On the other hand, alumina and water exhibit chemisorption whereas alumina and hydrogen do not; and alumina is a good catalyst for dehydration reactions but is ineffective for dehydrogenation. Non-refractory oxides like zinc oxide which can lose or gain oxygen from their lattices are good catalysts for vapour phase oxidation. It seems, therefore, that in the above examples, whenever chemisorption occurs, surface catalysis is possible. It is now generally accepted that chemisorption is a prerequisite for catalytic activity.

All surface catalysis reactions proceed by a series of steps as follows:

(*a*) Diffusion of reactant molecules to the surface
(*b*) Adsorption of reactant molecules, suitably orientated on the surface for reaction.
(*c*) Reaction between adsorbed molecules to give products.
(*d*) Desorption of products.
(*e*) Diffusion of products away from surface.

Step (*c*) possesses the highest energy barrier of the five steps in most reactions, so that this is usually the slow step. Only when the catalyst is extremely 'active' is the reaction rate determined either by the rate at which molecules diffuse to or away from the surface or by the rate at which molecules are adsorbed. In all cases, however, the experimental activation energy (E) is considerably lower than that for the uncatalysed reaction (see Table 8.2).

Not all the surface of a catalyst possesses strong adsorptive properties. Many facts concerning the poisoning of these catalysts suggest that there are certain *active sites* or *centres* where adsorption is exceedingly high compared with the rest of the surface. For instance, it is possible to have almost the entire surface of a catalyst covered with adsorbed gas without its catalytic power being affected. It can be imagined that these sites are actually the irregular 'bumps' on the solid surface such as crystal edges and grain boundaries. On a catalyst in a finely divided state there are many more active sites than on a solid lump.

The exact mechanisms of surface-catalysed reactions are very difficult to sort out in detail. One possible way in which such reactions may occur is illustrated in Fig. 8.3.

The ethylene and hydrogen molecules diffuse to the nickel surface where they are adsorbed. Reaction occurs between suitably orientated molecules such as those in Fig. 8.3. The electrons in the π bond of the ethylene molecule are used in the bonding of the molecule to the surface. The H—H bond is weakened when the hydrogen molecule is adsorbed. Interaction can then occur between the adsorbed molecules to give a molecule of ethane which is then desorbed.

Reactions catalysed by surfaces are more common than at first appears. Glass, for example, can be a very efficient catalyst and the kinetics of apparently homogeneous gas reactions carried out in glass vessels can be very misleading. Such reactions are, in fact, very much quicker in the presence of glass than, say, when the reaction vessel is made of silica. The classic work of Bodenstein on the decomposition of hydrogen iodide was hindered by this difficulty at the beginning of the investigation.

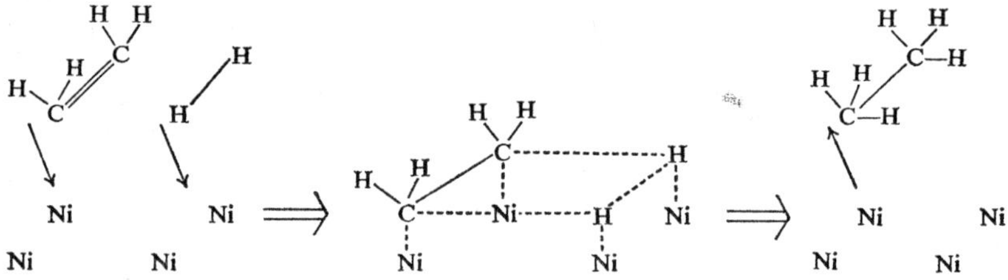

Fig. 8.3. Possible mechanism for the hydrogenation of ethylene on a nickel surface (for clarity the shape of the ethylene molecule has been slightly altered)

Enzyme catalysis

Experiment 8.5 Hydrolysis of starch catalysed by saliva

Procedure

Add 15 cm^3 of a 1 per cent suspension of starch in water to a boiling tube. Wash out the mouth with water so as to obtain a diluted solution of saliva. Add 15 cm^3 of this solution to the boiling tube. Place the tube in a water-bath at 50°C and at various intervals remove a few cm^3 for testing, making the first extraction after about 3 minutes. Add 1 drop of iodine solution to each sample removed (iodine gives a blue colour with starch and a red-orange colour with maltose). A control experiment without any saliva should be performed at the same time.

Discussion of results

The hydrolysis of starch to maltose is catalysed by α–amylase, an enzyme in saliva. This is one of the essential processes in the digestion of food in the human body. A further break-down of the molecules occurs when an intestinal enzyme (maltase) catalyses the formation of glucose from maltose.

Enzymes are proteins with very complex structures synthesized by the living cell. They act as catalysts in a wide range of biological processes and are found to possess remarkable activity. For instance, at 0°C only one molecule of the enzyme, *catalase*, is required to produce a rate of decomposition of hydrogen peroxide to water and oxygen of 4×10^4 molecules s^{-1} (the *turnover number* of catalase in this reaction). Experimental activation energies for these catalysed reactions are very low (see Table 8.2). They also possess remarkable specificity, some only functioning on one substance. The enzyme, urease (molecular weight about 480 000), will catalyse the hydrolysis of urea but will not do the same for N-methyl-urea, a very similar compound. All enzymes are effective at room or body temperature, and many are so sensitive to heat that they quickly become inactive at about 50°C. Many enzymes are sensitive to pH: in the digestion of proteins in the human body *pepsin* works at low pH in the stomach and *trypsin* under slightly acid to slightly alkaline conditions in the intestines.

Why are enzymes so specific in their catalytic action? There is no easy answer to this question. The concept of *active sites* has been used in enzyme chemistry as well as in surface catalysis. The enzyme surface can be considered to possess molecular indentations, or templates, into which only one type of molecule can 'fit'. It may be that an active site is a position on the enzyme molecule occupied by a certain functional group which can interact chemically with the reactant molecules. The great activity of enzyme catalysts can be ascribed to the very large number of active sites on the surfaces.

D. Some special features of catalytic action

Autocatalysis

In the reaction between permanganate ions and oxalic acid (or an oxalate) in acidic solution, the rate of reaction is very low until some of the products have been formed: the reaction is then accelerated.

$$2MnO^-_{4(aq)} + 16H^+_{(aq)} + 5H_2C_2O_{4(aq)} = 2Mn^{2+}_{(aq)} + 10CO_{2(g)} + 8H_2O_{(l)}$$

When manganese(II) ions are added to the *initial* reaction mixture, the reaction is fast from the beginning. This evidence suggests that the reaction is catalysed by one of the products, manganese(II) ions. This type of catalysis is known as *autocatalysis*. One further example of autocatalysis is described below, and another has already been investigated in Experiment 5.3(*b*).

Experiment 8.6 The bromination of acetone

$$Br_2 + CH_3COCH_3 = CH_3COCH_2Br + HBr$$

Procedure

Dissolve 5 drops of bromine liquid in about 10 cm^3 of tetrachloromethane. (Neither the solution nor the bromine liquid should be allowed to come into contact with the skin.) Add 1·0 cm^3 of this solution from a burette into 30 cm^3 of acetone in a boiling tube. Leave to stand for about a minute. Then place the tube in a water bath maintained at 60°C. Note the sudden colour change.

Cool the contents of the tube to room temperature and add a further 1·0 cm^3 of bromine solution from the burette. Close the tube with a bung, shake and note the time taken for the colour change to occur again, holding the tube against a piece of white paper. Then add 2·0 cm^3 of bromine solution, again noting the time for the colour change to occur.

Discussion of results

These results can be interpreted in terms of autocatalysis. In view of what is known about the iodination of acetone in acidic solution (Experiment 4.2), which of the products may be the autocatalyst? How could this idea be tested?

Increasing the efficiency of catalysts

Some substances, although having little or no catalytic activity themselves, can considerably increase the activity of catalysts to which they are added. For instance, it is sometimes found that enzymes contain metal ions which are essential for their activity: these are called *activators*. *Co-enzymes*, which are not proteins and have much simpler structures than enzymes, are sometimes responsible for the activity of certain enzymes. They may work with more than one enzyme and in several different reactions. Many enzymes use B vitamins as co-enzymes. In surface catalysis such substances are called *promoters*. The use of certain oxides as well as the usual iron catalyst in the Haber Process (see Table 8.4) is a well-known example.

The mechanism of promoter action is not understood. It could be that no one theory can cover all the examples. However, an important effect may be to introduce more points of strain into the lattice of the catalyst surface so that the number of active sites is increased.

Inhibitors and poisons

Inhibitors are substances which slow down the rate of a reaction. The term 'negative catalyst' which is sometimes used is really a misnomer. Such substances never act by providing *new* reaction pathways like true catalysts, but merely inhibit the progress of the reaction along an *existing* pathway.

When the inhibitor reduces the efficiency of a surface catalyst it is usually called a *catalyst poison*. Small amounts of arsenic compounds considerably reduce the activity of a platinum catalyst in the oxidation of sulphur dioxide to sulphur trioxide. A copper catalyst can be poisoned by carbon monoxide or mercury during the hydrogenation of ethylene. Such poisons probably act by competing with the reactants for the active sites on the surface of the catalyst. If the poison is adsorbed, the activity of the catalyst is reduced. The idea of active sites arose from a study of poisons because it was found that only minute quantities of poisons, far less than that required to cover the catalyst surface with a monatomic layer, are required for the catalyst to be rendered useless. In the hydrogenation of ethylene, for instance, only a small quantity of mercury is required to reduce the rate of reaction to less than 1 per cent of its normal rate.

Enzymes can also be poisoned easily. Poisons for these catalysts are the same kind of substances that poison surface catalysts (mercury, carbon monoxide, hydrogen sulphide, arsenic compounds, and so on). Because these substances inhibit vital biochemical processes catalysed by enzymes they are also strong *physiological* poisons.

There is a wide range of both catalysed and uncatalysed reactions which can be inhibited apart from those mentioned above. Five examples are:

(*a*) Nitrogen monoxide in the thermal decomposition of acetaldehyde vapour.
(*b*) Glycerine or dilute acids in the decomposition of hydrogen peroxide.
(*c*) Lead tetraethyl in the combustion of petrol.
(*d*) Nitrogen trichloride in the reaction between hydrogen and chlorine.
(*e*) Antoxidants such as phenolic compounds or derivatives of aromatic amines may be used either to prevent atmospheric oxidation of aqueous sulphite solutions or to reduce the formation of gums in fuels.

In these examples, the inhibitor may act by preventing one or more steps in the reaction mechanism from taking place. When the reaction concerned is a chain reaction (see page 141) the inhibitor is referred to as a 'chain breaker'. The thermal decomposition of acetaldehyde vapour is a chain reaction, and it seems that nitrogen monoxide, which possesses a single unpaired electron, reacts with the intermediate so inhibiting the reaction.

E. A comparison between catalytic effects and the effects of light on rates of reaction

Light affects the rate of many reactions: mixtures of hydrogen and chlorine do not react at a measurable rate in the dark at room temperature, but they can explode when exposed to sunlight; precipitates of the silver halides darken on exposure to light due to the formation of submicroscopic specks of silver in some grains (this is the basis of the use of silver bromide in photography); and the decomposition of hydrogen peroxide solutions is sensitive to light.

Why does light have such an effect on these, and other, reactions? A few reactant molecules can absorb visible light, and rather more can absorb ultra-violet radiation. Light of these wavelengths possesses sufficient energy to break strong bonds and so to *initiate* a reaction. Reactive intermediates are so produced which can react to form the products. Reactions which are not light-sensitive do not have reactants capable of absorbing light of an appropriate wavelength.

A careful distinction should be made between the effect of light and the effect of a catalyst on the rate of reaction: light can initiate a reaction but is consumed during the reaction; but a catalyst provides a new reaction pathway without being consumed during the reaction. Moreover, in some reversible reactions (see Exercise 1) only the forward reaction is sensitive to light, whereas a catalyst always affects the forward and reverse reactions equally.

Questions and exercises

1. Investigate the effect of light on the reaction between iron(III) ions and thionine. (See *S.S.R.*, 1964, **157**, 680).
2. Look up the details of the use of benzoyl peroxide in the polymerization of styrene or vinyl chloride. Is it best described as an initiator or a catalyst?
3. Investigate the effect of light from burning magnesium on freshly precipitated silver chloride and silver bromide.
4. The decomposition of hydrogen peroxide can be catalysed by iodide ions.

$$2H_2O_{2(aq)} = 2H_2O_{(l)} + O_{2(g)}$$

The experimental rate expression is:

$$-\frac{1}{2}\frac{d[H_2O_2]}{dt} = \text{Const.}[H_2O_2][I^-]$$

Suggest a possible mechanism involving the catalyst.

5. Investigate the kinetics of the 'blue bottle' experiment. Find the time taken for the blueing–deblueing cycle to occur for different concentrations of each reactant.

Quantities:	NaOH g	*Glucose g*	*Water* cm³	*Methylene blue solution (no. of drops)*
	1·0	2·0	100	20
	1·0	2·0	100	40
	1·0	4·0	100	20
	2·0	4·0	100	20

The concentration of methylene blue solution is as in Experiment 8.1. Use the results to suggest an approximate rate expression.

6. Investigate the effect of iron(II) and copper(II) ions on the rate of the reaction:

$$2I^-_{(aq)} + S_2O^{2-}_{8(aq)} = I_{2(aq)} + 2SO^{2-}_{4(aq)}$$

Use about 0·5 M potassium iodide solution and about 0·02 M potassium persulphate solution, and add a few drops of 0·005 M iron(II) ammonium sulphate or copper(II) sulphate solutions. Devise a suitable clock method of estimating the reaction rate.

9 The hydrolysis of alkyl halides and esters

The decomposition of a substance by reaction with water is called *hydrolysis* and this definition includes those reactions which are carried out in acid and alkaline aqueous solutions (acid hydrolysis and alkaline hydrolysis respectively). The organic reactions investigated below are the hydrolyses of alkyl halides and esters. The results of experiments like these have provided organic chemists with some important clues to possible mechanisms for these reactions.

The hydrolysis of alkyl halides

A. *Iodoethane*

Experiments show that the hydrolysis is slow in neutral or acidic solutions [equation (1)] but is more rapid in alkaline solution [equation (2)].

$$C_2H_5I_{(aq)} + H_2O_{(l)} = C_2H_5OH_{(aq)} + H^+_{(aq)} + I^-_{(aq)} \quad (1)$$

$$C_2H_5I_{(aq)} + OH^-_{(aq)} = C_2H_5OH_{(aq)} + I^-_{(aq)} \quad (2)$$

The rate of the reaction in alkaline solution [equation (2)] is measured in Experiment 9.1 and an attempt is made to determine the order of the reaction.

Experiment 9.1 The alkaline hydrolysis of iodoethane
Because iodoethane is only sparingly soluble in water, the reaction is carried out in a mixture of acetone and water. A temperature of about 50°C is required to give a suitable rate of reaction. The rate is determined by measuring the hydroxide ion concentration at several times during the reaction, and from the stoichiometry of equation (2):

$$-\frac{d[OH^-]}{dt} = -\frac{d[C_2H_5I]}{dt}$$

Samples of the reaction mixture are withdrawn at known times and are titrated with standard acid.

The presence of carbonate ions would cause the end-point to fade and be indistinct, and since high accuracy is required in this experiment it is therefore strongly recommended that carbonate-free reagents should be used throughout.† (Standard carbonate-free sodium hydroxide and hydrochloric acid solutions may be purchased from suppliers of chemical reagents.) Alternatively, the sodium hydroxide solution may be prepared from a saturated solution which has been filtered (sodium carbonate is insoluble in concentrated caustic soda), and the acid may be prepared by diluting concentrated acid with carbonate-free water. Carbon dioxide may be removed from water by boiling. The carbon dioxide-free reagents should be stored in closed containers.

† Results of greater accuracy could be obtained by preparing a larger volume of reaction mixture than is suggested, and taking larger, say 25 cm^3, samples for titration.

The results can be interpreted most easily when the iodoethane and sodium hydroxide are present in equimolar concentrations in the reactant mixture. Because these substances react in equimolar proportions [see equation (2)] their concentrations then remain equal throughout the reaction.

Reagents

For the three experiments involved (see later) the reagents required are:

150 cm^3 1·00 M sodium hydroxide, 200 cm^3 1·00 M hydrochloric acid and about 3 dm^3 water, all carbonate-free; iodoethane (about 12 g) and acetone (150 cm^3).

Procedure

Prepare 100 cm^3 0·200 M sodium hydroxide solution in water and 100 cm^3 0·200 M iodoethane solution in acetone. Stand the flasks containing the solutions in a thermostat bath maintained at 50°C for at least 5 minutes. Then thoroughly mix the reactants in one of the flasks. To prevent loss by evaporation, the reaction vessel should be closed with a bung except when a sample is being taken. After thermal equilibrium has been attained, use a pipette fitted with a pipette bulb, to remove a 10 cm^3 sample of the reaction mixture. Discharge the sample into about 20 cm^3 water at room temperature, noting the time when the pipette is half empty. Titrate with 0·0500 M hydrochloric acid using methyl red as indicator. Repeat the sampling and titration at about 5- to 7-minute intervals for about 1 hour. The readings should be tabulated as in Table 9.1.

Adding the sample to water slows the reaction (dilution and cooling) but does not stop it, so that the titration of the samples should be performed immediately they have been diluted. The sampling pipette should be rinsed with water after a sample is taken and left to drain.

Table 9.1. Alkaline hydrolysis of iodoethane

$[OH^-]_0 = [C_2H_5I]_0 = 0{\cdot}10\ mol\,dm^{-3}$

Time† t min	*Titre* $V_t\,cm^3$
0	20·25
15	19·50
27	18·90
38	18·50
49	18·10
60	17·80

Discussion of results

Suppose V_t cm^3 acid reacts with the alkali in the sample taken at time t. Then at time t:

$$[OH^-]_t = cV_t \tag{3}$$

† The zero on the time scale does not affect the results. The clock was started when the first sample was diluted.

where c is the constant of proportionality which relates volume of acid to concentration of hydroxide ions in mol dm^{-3}.

If the reaction is first order, the integrated rate expression is

$$\log_{10}[OH^-]_t = -\frac{k_1 t}{2{\cdot}30} + \log_{10}[OH^-]_0 \qquad \text{(see Table 4.1)}$$

(The rate expression could be written in terms of $[C_2H_5I]$.) Substitution from equation (3) gives

$$\log_{10} V_t = -\frac{k_1 t}{2{\cdot}30} + \log_{10} V_0 \tag{4}$$

where V_0 is the volume of acid which would have been necessary for a sample taken when $t = 0$.

If the reaction is second order, the integrated rate expression is

$$\frac{1}{[OH^-]_t} = k_2 t + \frac{1}{[OH^-]_0} \qquad \text{(see Table 4.1)}$$

Substitution from equation (3) gives

$$\frac{1}{V_t} = ck_2 t + \frac{1}{V_0} \tag{5}$$

Thus if the reaction is first order, a plot of $\log_{10} V_t$ against t will give a straight line, whereas if it is second order, a plot of $1/V_t$ against t will give a straight line. Some results are given in Table 9.1 and the graphs are shown in Fig. 9.1.

It is immediately apparent that straight lines can be drawn through the points for both plots. Can this indicate that the reaction is both first and second order?

If V_t is plotted against t, it is found that although there is a trend indicating curvature, the scatter of the points away from the 'best straight line' is only slightly greater than would be expected from experimental error. Can this mean that the reaction is zero order?

For complete reaction, V_t would decrease by about 20 cm^3 to zero. Table 9.1 shows that V_t decreases by nearly $2\frac{1}{2}$ cm^3, so that only about $2\frac{1}{2}/20 \times 100 = 12\frac{1}{2}$ per cent of the complete reaction has occurred in the hour during which readings were taken. In Chapter 4 it was stated that at least 50 per cent of a reaction should be studied if its order is to be established unambiguously. This example shows clearly what can happen when readings cover only a small fraction of the total reaction. Either the reaction can be followed for a much longer time or, more conveniently, the effect of changing the initial concentrations of the reactants can be examined.

Suppose that the reaction is second order, so that the rate expression may be written

$$-\frac{d[OH^-]}{dt} = k_2[OH^-]^2$$

The slope of the second order plot (Fig. 9.1) is equal to ck_2 [equation (5)] and this slope should not change if the initial concentrations of the reactants are altered.

However, the data have also been interpreted as fitting the first order rate expression:

$$-\frac{d[OH^-]}{dt} = k_1[OH^-]$$

Thus the slope of the first order plot, which was meant to give $-k_1/2{\cdot}30$, is related to $-k_2[OH^-]/2{\cdot}30$, and this slope will therefore change if the initial concentrations of the reactants are altered. In particular, if the reaction is second order and Experiment 9.1 is repeated with the initial reactant concentrations doubled, the slope of the V_t against t plot will be quadrupled; the slope of the first order plot will be doubled; and the slope of the second order plot will be unchanged. If the reaction is first order, and Experiment 9.1 is repeated with the initial concentrations doubled, the slope of the V_t against t plot will be doubled; the slope

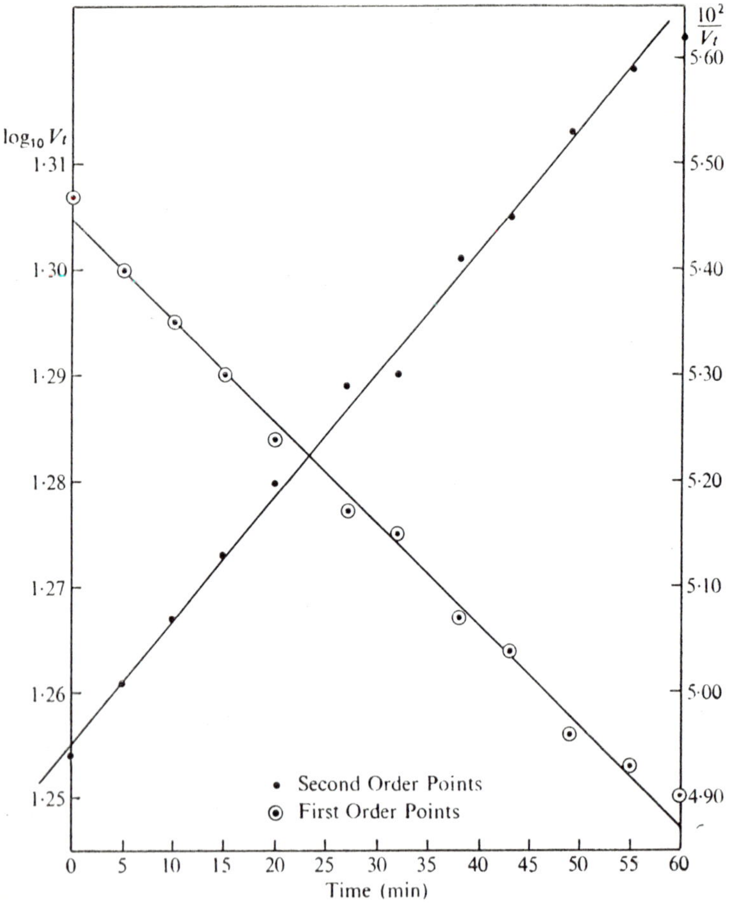

Fig. 9.1. First and second order plots for the alkaline hydrolysis of iodoethane. Temp. 50°C; Solvent, 1:1 water and acetone by volume. Initial concentrations of OH^- and C_2H_5I, 0·10 M

of the first order plot will be unchanged; and the slope of the second order plot will be halved.

If this procedure is followed, it is important that the concentration of acid used in all the experiments should be constant, otherwise the proportionality constant c will be altered, and this will interfere with the comparison of the slopes.

Determination of the order of the reaction

Experiment 9.1 was repeated using halved and then doubled initial reactant concentrations. The results of these experiments are contained in Table 9.2. First and second order plots were drawn as in Figure 9.1, and the slopes of these lines were determined and are recorded in Table 9.3.

Table 9.2. Alkaline hydrolysis of iodoethane

$[OH^-]_0 = [C_2H_5I]_0 = 0{\cdot}050$ mol dm^{-3}		$[OH^-]_0 = [C_2H_5I]_0 = 0{\cdot}20$ mol dm^{-3}	
Time t min	Titre V_t cm^3	Time t min	Titre V_t cm^3
0	10·25	0	39·90
5	10·10	8	38·80
15	9·95	17	37·25
40	9·70	34	35·00
60	9·50	50	33·30

Table 9.3. Slopes of first and second order plots for the alkaline hydrolysis of ethyl iodide

$[OH^-]_0 = [C_2H_5I]_0$ (mol dm^{-3})	*First order slope* (min^{-1})	*Second order slope* (cm^{-3} min^{-1})
0·20	$-1{\cdot}60 \times 10^{-3}$	$1{\cdot}05 \times 10^{-4}$
0·10	$-0{\cdot}95 \times 10^{-3}$	$1{\cdot}15 \times 10^{-4}$
0·050	$-0{\cdot}45 \times 10^{-3}$	$1{\cdot}10 \times 10^{-4}$

The second order slopes $(1{\cdot}10 \pm 0{\cdot}05) \times 10^{-4}$ cm^{-3} min^{-1} vary by less than 5 per cent, whereas the first order slopes are very approximately doubled when the initial reactant concentration is doubled.

These results show that the reaction is second order, a likely rate expression being

$$-\frac{d[C_2H_5I]}{dt} = -\frac{d[OH^-]}{dt} = k_2[C_2H_5I][OH^-] \qquad (6)$$

However, other rate expressions which are consistent with the experimental results are

$$-\frac{d[OH^-]}{dt} \propto [OH^-]^2 \qquad (7a)$$

$$-\frac{d[OH^-]}{dt} \propto [C_2H_5I]^2 \qquad (7b)$$

It can be shown that equation (6) is correct from the results of experiments in which the initial concentrations of the reactants are not equal.

B. 2-*Chloro-2-methylpropane* (*tertiary butyl chloride*)

The procedure for the initial investigation of this reaction is similar to that employed in Experiment 9.1, but some modifications are necessary. The rate of hydrolysis is slow in a solvent consisting of equal volumes of water and acetone, but it is found that a suitable rate at 25°C is obtained by using equal volumes of water and ethanol as the solvent. When samples are taken, the rate of reaction is decreased by diluting them with acetone. It is not necessary to use carbonate-free reagents and water because the reaction is more rapid than that in Experiment 9.1 and there is a large difference between the initial and final titres.

Experiment 9.2 The alkaline hydrolysis of 2-chloro-2-methylpropane

$$(CH_3)_3CCl_{(aq)} + OH^-_{(aq)} = (CH_3)_3COH_{(aq)} + Cl^-_{(aq)} \qquad (8)$$

Reagents

250 cm³ of a solution of 2-chloro-2-methylpropane in ethanol. The molarity of this solution should be known to an accuracy of about 10 per cent and should be between 0·1 and 0·2 M.

250 cm³ of sodium hydroxide solution in water with the same molarity as the alkyl halide solution.

250 cm³ hydrochloric acid with a molarity about, or slightly less than, one half of the molarity of the sodium hydroxide solution.

Procedure

Stand the flasks containing the reactant solutions, and a 600 cm³ or larger flask, in a thermostat bath maintained at 25°C and allow thermal equilibrium to be attained. Add the solutions to the flask, mix them well and allow them to stand for about 5 minutes in the bath. The flask should be closed with a bung except when a sample is being taken. Using a pipette fitted with a pipette bulb, withdraw a 25 cm³ sample of the reaction mixture and discharge it into about 40 cm³ acetone, noting the time when the pipette is half empty. Dilution with acetone slows down, but does not stop, the reaction, so that the mixture should be titrated immediately with hydrochloric acid, using methyl red as the indicator.

The end-point fades due to the presence of carbonate in the reaction mixture, and it is necessary to standardize the titration procedure. It is suggested that the end-point should be taken when just sufficient acid has been added for the red colour to persist for 10 seconds.

Repeat the withdrawal and titration of samples at about 5- to 7-minute intervals for about 1 hour, rinsing the pipette with water after taking a sample.

Determination of the order of the reaction

The treatment of the results is similar to that described for Experiment 9.1. The results of an experiment are given in Table 9.4 and the first and second order plots are shown in Fig. 9.2.

Table 9.4. Alkaline hydrolysis of 2-chloro-2-methylpropane

$[OH^-]_0 = [(CH_3)_3CCl]_0 = 0{\cdot}080\ mol\,dm^{-3}$

Time t min	*Titre* V_t cm³
0	22·73
7	18·50
15	15·45
27	11·55
44	8·00
60	5·55

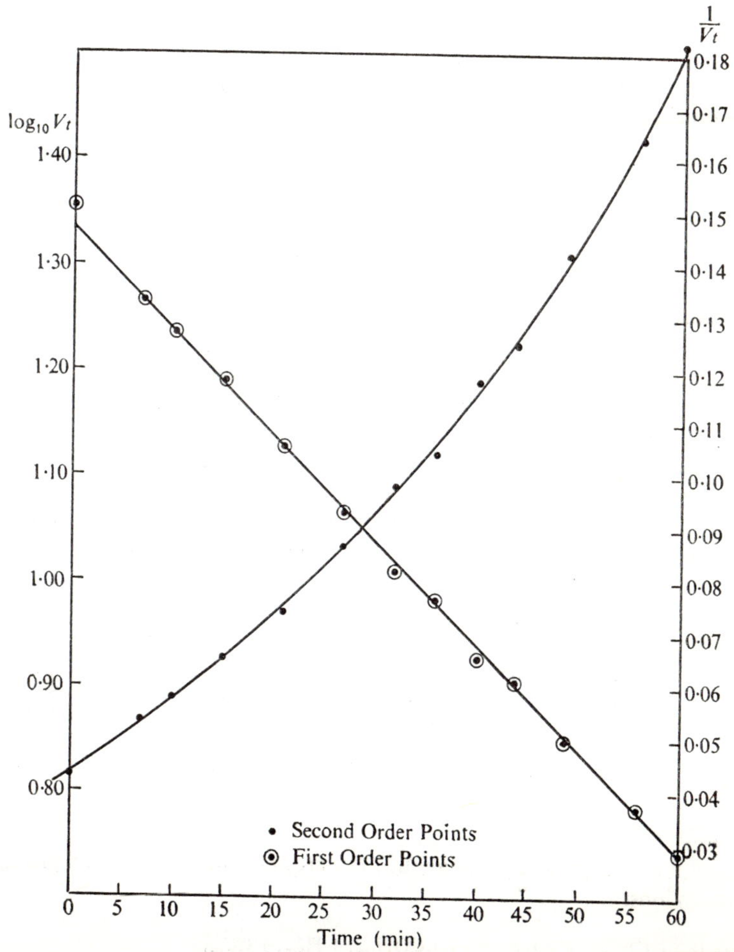

Fig. 9.2. First and second order plots for the alkaline hydrolysis of 2-chloro-2-methylpropane

First order slope = $-9{\cdot}9 \times 10^{-3}$ min^{-1}

First order rate constant = 0·023 min^{-1}

V_t decreases by about 17 cm^3 whereas for complete reaction, V_t would decrease to zero. Thus the experimental results cover about 75 per cent of the reaction.

The reaction is first order and the simplest rate expressions which are consistent with this result are

$$-\frac{d[OH^-]}{dt} = k_1[OH^-] \quad (9)$$

and

$$-\frac{d[OH^-]}{dt} = k_1[(CH_3)_3CCl] \quad (10)$$

For equation (9) to hold, the reaction is zero order in alkyl halide, whereas if equation (10) applies, the reaction is zero order in hydroxide ions. Other rate expressions such as

$$-\frac{d[OH^-]}{dt} = k_1\frac{[OH^-]^2}{[(CH_3)_3CCl]} \quad (11)$$

are also consistent with the experimental results.

Determination of the order with respect to each reagent

The results of Experiment 9.2 give the overall order of the reaction, but because the reactant concentrations were equal, the mathematical treatment of the results cannot distinguish between equations (9) and (10) and others. For this, unequal reactant concentrations are necessary.

If equation (10) applies, the rate of hydrolysis is independent of the hydroxide ion concentration. If this holds even at very low hydroxide ion concentrations, the rate of hydrolysis in alkaline solution may be equal to the rates in neutral or even acid solutions. Experiment 9.3 is designed to test this possibility.

Experiment 9.3

Prepare 50 cm^3 of an approximately 0·1 M solution of 2-chloro-2-methylpropane in a solvent of 50 per cent water and 50 per cent ethanol by volume. Test the pH of the reaction mixture every 10 or 20 seconds with indicator paper. At first, use the pH 4–6 paper, and later on the pH 1–4 paper. After about 5 minutes test the reaction mixture less frequently, say every 2 minutes, for about 15 minutes.

Discussion of results

The pH of the mixture begins at about 7 but drops because the reaction produces hydrogen ions.

$$(CH_3)_3CCl_{(aq)} + H_2O_{(l)} = (CH_3)_3COH_{(aq)} + H^+_{(aq)} + Cl^-_{(aq)} \quad (12)$$

The final concentration of hydrogen ions should be about 10^{-1} M (i.e. pH = 1). At pH = 2 when the hydrogen ion concentration is 10^{-2} M, the extent of the reaction is $10^{-2}/10^{-1} \times 10^2$ or 10 per cent. This experiment shows that it does not take long for 10 per cent of the reaction to be completed. Therefore, the hydrolysis of 2-chloro-2-methylpropane in an initially neutral solution occurs at an easily measurable rate. This argues that the rate of hydrolysis *may* not depend on $[OH^-]$, and that equation (10) appears to be a possible rate expression.

Experiment 9.4 tests equation (10) quantitatively by allowing the hydrolysis to occur in an initially neutral solution and by estimating the concentration of hydrogen ions produced at known times.

Experiment 9.4

Procedure

Stand a flask containing about 250 cm^3 of a mixture of equal volumes of ethanol and water in a thermostat bath at 25°C until thermal equilibrium is attained. Add about 2 g of 2-chloro-2-methylpropane and mix well. After about 5 minutes (as in Experiment 9.2), withdraw a 25 cm^3 (or 20 cm^3) sample of the reaction mixture, run it into acetone and titrate it immediately with approximately decimolar sodium hydroxide solution using methyl red as the indicator. Repeat the sampling and titration at about 6-minute intervals for about 1 hour. The end-point does not fade in these titrations as there can be no carbonate ions in the acid reaction mixture.

Allow the reaction mixture to stand overnight and then determine V_∞, the volume of the sodium hydroxide solution required to react with the acid in a 25 cm^3 (or 20 cm^3) sample when the reaction is complete. At least two titrations should be performed. If the molarities of the alkyl halide and alkali solutions are known accurately, V_∞ can be calculated.

Determination of the order of the reaction

If the reaction is first order, the integrated rate equation is:

$$\log_{10}[(CH_3)_3CCl]_t = -\frac{k_1}{2{\cdot}30}.t + \log_{10}[(CH_3)_3CCl]_0 \qquad (13)$$

Now

$$[(CH_3)_3CCl]_0 \propto V_\infty$$
$$[(CH_3)_3CCl]_t \propto V_\infty - V_t$$

Substitution in equation (13) gives

$$\log_{10}(V_\infty - V_t) = -\frac{k_1}{2{\cdot}30}.t + \log_{10} V_\infty$$

where the necessary proportionality constant disappears. Thus a graph of $\log_{10}(V_\infty - V_t)$ against t will give a straight line if the reaction is first order with respect to the alkyl halide.

Some results are given in Table 9.5 and the first order plot is a straight line. The reaction is thus first order in 2-chloro-2-methylpropane, and the rate expression is:

$$-\frac{d[(CH_3)_3CCl]}{dt} = k_1[(CH_3)_3CCl] \qquad (10)$$

Table 9.5. Hydrolysis of 2-chloro-2-methylpropane in an initially neutral solution (Experiment 9.4)

Time t min	V_t cm^3	
3	2·1	
11	6·4	
18	9·4	(20·0 cm^3 samples taken)
28	13·1	
45	18·0	
56	20·6	$V_\infty = 36{\cdot}0$ cm^3 by titration

A comparison of the values of k_1 obtained in Experiments 9.2 and 9.4 (Table 9.6) shows that they are well within a factor of 2 of one another. Because the alkaline and aqueous hydrolyses have identical rate expressions and similar rate constants, it is possible that these reactions have a common rate determining step.

Table 9.6. Rate constants for Experiments 9.2 and 9.4

Experiment	*Slope* min^{-1}	k_1 min^{-1}
9·2	$-9{\cdot}9 \times 10^{-3}$	$2{\cdot}3 \times 10^{-2}$
9·4	$-6{\cdot}5 \times 10^{-3}$	$1{\cdot}5 \times 10^{-2}$

The mechanisms for the hydrolysis of iodoethane and 2-chloro-2-methylpropane

The rate expression for the alkaline hydrolysis of iodoethane is

$$-\frac{d[C_2H_5I]}{dt} = k_2[C_2H_5I][OH^-] \qquad (6)$$

This suggests that the rate determining step involves one hydroxide ion and one iodoethane molecule, and since these are the reactant proportions in the stoichiometric equation it can be postulated that the reaction is elementary.

$$OH^- + C_2H_5I \rightarrow C_2H_5OH + I^- \qquad (2)$$

The low rate of hydrolysis in neutral and acid solutions is then explained by the very low concentrations of hydroxide ions in these solutions. However, there is the possibility that the mechanism could be different in these solutions and other experiments would be needed to test mechanisms such as

$$C_2H_5I + H_2O \rightarrow C_2H_5OH + I^- + H^+$$

The Transition State Theory (see page 93) would explain mechanism (2) as the simultaneous breaking of the C—I bond and making of the O—C bond. The activated complex could therefore be represented as

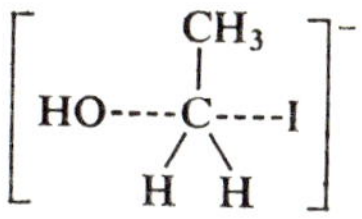

in which the broken lines represent bonds which are being made or broken and which are weaker than the normal covalent bonds. The course of the reaction would then be:

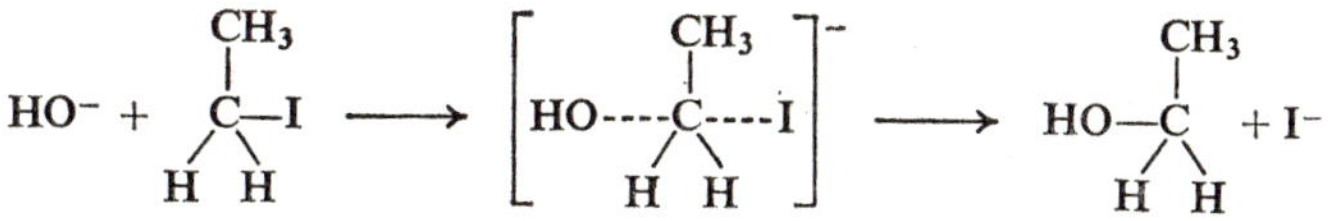

The rate expression for the hydrolysis of 2-chloro-2-methylpropane under the conditions of Experiment 9.4 is:

$$-\frac{d[(CH_3)_3CCl]}{dt} = k_1[(CH_3)_3CCl] \qquad (10)$$

However, the variation of water concentration during the reaction is very small compared with the change of the alkyl halide concentration. It is therefore possible that the rate expression is:

$$-\frac{d[(CH_3)_3CCl]}{dt} = k_2[(CH_3)_3CCl][H_2O] \qquad (14)$$

where $k_1 = k_2[H_2O]$.

For rate expression (10) a slow step involving the decomposition of one molecule of the alkyl halide can be postulated. If the rate determining step is written as the first step the mechanism could be:

Mechanism 1 (general)

$(CH_3)_3CCl$ = intermediate(s) Step 1 (slow)

intermediate(s) + $H_2O \rightarrow (CH_3)_3COH + H^+ + Cl^-$ Step 2 (fast)

Addition of these two steps gives the stoichiometric equation for the hydrolysis by water.

For rate expression (14) the reaction could be postulated to be elementary, i.e.

Mechanism 2

$$(CH_3)_3CCl + H_2O \rightarrow (CH_3)_3COH + H^+ + Cl^- \qquad (12)$$

The choice is therefore between a unimolecular rate determining step in a complex reaction and a simpler bimolecular reaction. Other information is required if a choice is to be made between these two mechanisms.

Analysis of the reaction mixture after completion of the reaction shows that there are several products whose proportions vary with the reaction conditions (temperature, nature of solvent and so on). The main products are:

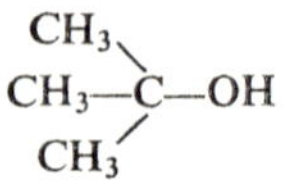

2-hydroxy-2-methylpropane
(as expected)

(I)

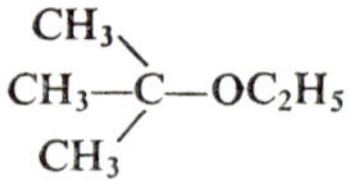

2-ethoxy-2-methylpropane

(II)

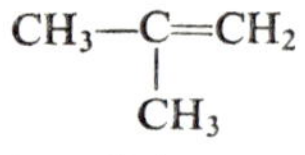

2-methylpropene

(III)

The formation of product II could be explained by a step such as

$$(CH_3)_3CCl + C_2H_5OH \rightarrow (CH_3)_3COC_2H_5 + H^+ + Cl^-$$

which is similar to equation (12). However, the formation of III is difficult to explain in terms of mechanism 2. Can mechanism 1 explain the presence of III?

Firstly it must be decided what is the likely structure of the intermediate in mechanism 1. Since the $(CH_3)_3C$ group is present in the products as well as in the reactants it seems likely that it is present in the intermediate. There are then two possibilities for step 1.

$$(CH_3)_3C{:}Cl = (CH_3)_3C^+ + Cl^-$$

$$(CH_3)_3C{:}Cl = (CH_3)_3C\cdot + Cl\cdot$$

In each case the C—Cl bond is broken. When the ions are formed both electrons are kept by the chlorine atom giving it an excess charge of –1. When the free radicals are formed the bond is broken so that one electron is retained by the $(CH_3)_3C$ group and one by the chlorine atom.

Both $(CH_3)_3C^+$ and $(CH_3)_3C\cdot$ are likely to be very unstable with respect to $(CH_3)_3CCl$ but the formation of the cation is more likely in an aqueous medium because of hydration. Moreover, the nature of the solvent is important in determining the rate of the reaction. Hydrolysis occurs about 30 000 times faster in 50 per cent aqueous ethanol than in 100 per cent ethanol, and the rate in 100 per cent acetone is slower still. The suggestion could be made that the more polar the solvent the faster is the hydrolysis. It seems that the ionic mechanism is more likely than the free radical mechanism.

It is now possible to explain the formation of all three products I to III using mechanism 1 and the $(CH_3)_3C^+$ intermediate. Mechanism 1, therefore, can explain the experimental evidence better than mechanism 2.

Mechanism 1 (detailed)

$$(CH_3)_3CCl \rightarrow (CH_3)_3C^+ + Cl^- \qquad \text{Step 1 slow}$$

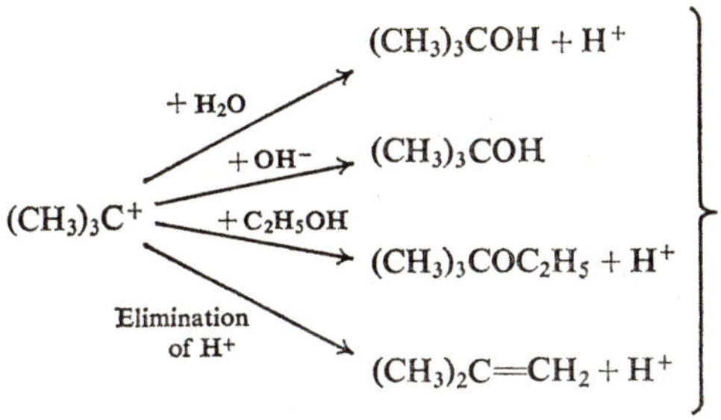

Alternative second steps, all faster than Step 1.

Other evidence for the postulated mechanisms of the hydrolysis of iodoethane and 2-chloro-2-methylpropane comes from an investigation of the stereochemistry of the reactants and products. In general, the primary alkyl halides are found to undergo hydrolysis by second order reactions (which are interpreted as involving a single bimolecular step), whereas the branched chain alkyl halides undergo hydrolysis by first order reactions (which are thought to involve a unimolecular rate determining step). The differences in behaviour have been explained in terms of the differences in the structure of the molecules, but discussion of this falls outside the scope of this book.†

The hydrolysis of esters

Esters react only slowly with water and the reaction is reversible. If R and R′ represent alkyl groups the general equation for the reaction is

$$RCOOR' + H_2O \rightleftharpoons RCOOH + R'OH \qquad (15)$$

Hydrolysis is usually carried out in either acid or alkaline solution because these reactions are more rapid than those in neutral solution. The alkaline hydrolysis is frequently termed saponification.

Experiment 9.5 The saponification of ethyl acetate

In alkaline solution the equation for the reaction is:

$$CH_3COOC_2H_{5(aq)} + OH^-_{(aq)} = CH_3COO^-_{(aq)} + C_2H_5OH_{(aq)} \qquad (16)$$

Whereas the reaction in equation (15) is reversible, this reaction, equation (16), is irreversible.

(*a*) The overall order of the reaction can be found in experiments in which the concentrations of ester and alkali are equal.

† The reader should consult a standard reference book on organic reaction mechanisms such as *Mechanism in Organic Chemistry* by P. Sykes, for this information.

Reagents (all carbonate-free):

250 cm^3 0·05 M sodium hydroxide	in water
250 cm^3 0·05 M ethyl acetate	in water
250 cm^3 0·025 M hydrochloric acid	in water
ice	

Procedure

Stand the flasks containing the reactants in a thermostat bath at 20°C. After thermal equilibrium has been attained mix the reactants and again allow thermal equilibrium to be attained. Then withdraw a 25 cm^3 sample. Discharge it into about 30 cm^3 of ice-cooled water, noting the time when the pipette is half empty. Immediately titrate the diluted sample with 0·025 M hydrochloric acid using phenolphthalein as the indicator. Repeat the sampling and titration at about 2-minute intervals for the first three readings and then at about 4-minute intervals. Take readings for a total time of about 40 minutes.

Determination of the order of the reaction

If, at time t, V_t cm^3 of acid are needed to neutralize the alkali in the sample, the concentrations of ester and hydroxide ions present are equal and proportional to V_t. If therefore the reaction is first order, a plot of $\log_{10} V_t$ against t will give a straight line, whereas if it is second order, a plot of $1/V_t$ against t will be linear.

Discussion of results

The overall order of the reaction is two and a possible rate expression is

$$-\frac{d[OH^-]}{dt} \propto [OH^-][CH_3COOC_2H_5] \qquad (17)$$

but other rate expressions which are consistent with the experimental results are

$$-\frac{d[OH^-]}{dt} \propto [OH^-]^2 \qquad (18)$$

and

$$-\frac{d[OH^-]}{dt} \propto [CH_3COOC_2H_5]^2 \qquad (19)$$

In previous similar examples, the correct rate expression was found by determining the order of the reaction with respect to one reactant only. Experiments were performed in which the concentration of one reagent was made effectively constant by using it in a large excess. However, in this example, a large excess of sodium hydroxide cannot be used because this is the reagent which is being estimated by titration, and a large excess of ester cannot be used because it is only slightly soluble in water.

(*b*) There are two procedures which will distinguish between the possible rate expressions in this reaction.

1. Repeat Experiment 9.5 but use unequal concentrations of ester and alkali. For example, the reactant solutions could be 250 cm^3

of 0·050 M ester and 250 cm^3 of 0·100 M sodium hydroxide. From the titrations, the concentrations of ester and alkali are calculated at several times during the reaction. If the reaction is first order with respect to each reagent, a plot of

$$\log_{10}\frac{\text{concentration of alkali at time } t}{\text{concentration of ester at time } t}$$

against t should be a straight line (see Table 4.1).

2. Repeat Experiment 9.5 using double the concentration of sodium hydroxide solution but keeping the same concentration of ester. Follow the reaction for long enough to obtain sufficient points for plotting the initial part of a V_t against t curve. Estimate the initial rate by drawing the tangent at $t = 0$ to this curve. Compare this initial rate with the one for Experiment 9.5 and deduce the order with respect to hydroxide ions.

In view of the results, which of the equations (17) to (19) is likely to be the correct rate expression?

Experiment 9.6 The acid hydrolysis of methyl acetate

$$CH_3COOCH_{3(aq)} + H_2O_{(l)} \rightleftharpoons CH_3COOH_{(aq)} + CH_3OH_{(aq)}$$

Reagents: 250 cm^3 1·0 M hydrochloric acid solution in water, in a flask fitted with a bung
500 cm^3 0·50 M sodium hydroxide solution
About 15 cm^3 methyl acetate in a small flask fitted with a bung

Procedure

Clamp the flasks containing the reactants in a thermostat bath at 40°C and allow thermal equilibrium to be attained. Add the methyl acetate to the hydrochloric acid and mix the reactants thoroughly. Start the clock and return the reaction mixture to the thermostat bath. After about three minutes withdraw a 10 cm^3 sample and run it into about 50 cm^3 water, noting the time when the pipette is half empty. The cooling and dilution slow down the reaction, but the sample should be titrated immediately with the 0·50 M sodium hydroxide solution, using phenolphthalein as indicator. Take samples at about 5-minute intervals for one hour, quenching and titrating as above.

Take several 10 cm^3 samples and titrate them as above, at least three hours after the start of the reaction, in order to determine V_∞.

Determination of the order of the reaction

The titre V_t at time t estimates both the hydrochloric acid which is present as the catalyst and the acetic acid which has been produced by the hydrolysis of the ester at that stage of the reaction. The titre V_∞ estimates the acid which is present as the catalyst and the acetic acid produced by complete hydrolysis. Since the concentration of the hydrochloric acid remains constant throughout the reaction, $(V_\infty - V_t)$ is proportional to the concentration of the ester in the reaction mixture at time t. Experiment shows that the observed value of V_∞ is within about 10 per cent of that expected for complete hydrolysis of the ester. This error in V_∞ does not affect the linearity of any plot, but does affect the calculation of a rate constant. First and second order plots can be drawn, $\log_{10}(V_\infty - V_t)$ against t and $1/(V_\infty - V_t)$ against t respectively, and the order of the reaction determined.

In Experiment 9.6, the concentration of the hydrogen ions from the hydrochloric acid remains constant throughout the reaction and the concentration of water remains effectively constant, so that the order calculated is the order with respect to methyl acetate.

This reaction is potentially autocatalytic due to the formation of acetic acid. However, this is a weak acid and the concentration of hydrogen ions produced from it is small compared with that from the hydrochloric acid.

Determination of the order with respect to the acid

The order with respect to the acid can be obtained by repeating Experiment 9.6 using a different concentration of acid, and comparing the values of the rate constants. If the reaction is first order with respect to the hydrogen ion concentration, halving the acid concentration should halve the rate constant.

If the experiment is repeated with the acid concentration halved, the same sodium hydroxide concentration can be used, but if it is doubled, a more concentrated alkali solution is required and the accuracy with which the ester concentration is estimated is thus decreased.

The mechanisms of the acid and alkaline hydrolyses of esters

The rate expression for the saponification of ethyl acetate was found to be

$$-\frac{d[OH^-]}{dt} = -\frac{d[CH_3COOC_2H_5]}{dt} = k_2[OH^-][CH_3COOC_2H_5]$$

which suggests that the rate determining step in the reaction involves one hydroxide ion and one ester molecule. These proportions are the same as in the stoichiometric equation for the reaction, so that at first sight it appears that the reaction is elementary, with a mechanism similar to that for the alkaline hydrolysis of iodoethane.

Mechanism 1

$$CH_3COOC_2H_5 + OH^- \longrightarrow \text{activated complex} \longrightarrow CH_3COO^- + C_2H_5OH$$

or

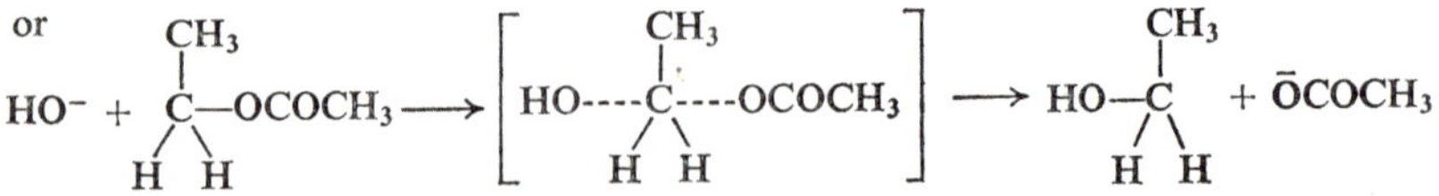

Broken lines represent bonds which are being either broken or made.

However, another mechanism is possible which is consistent with the observed rate expression. The hydroxide ion could attack the carbon atom in the carbonyl group so that the ester is cleaved in a different place from that in mechanism 1.

Mechanism 2

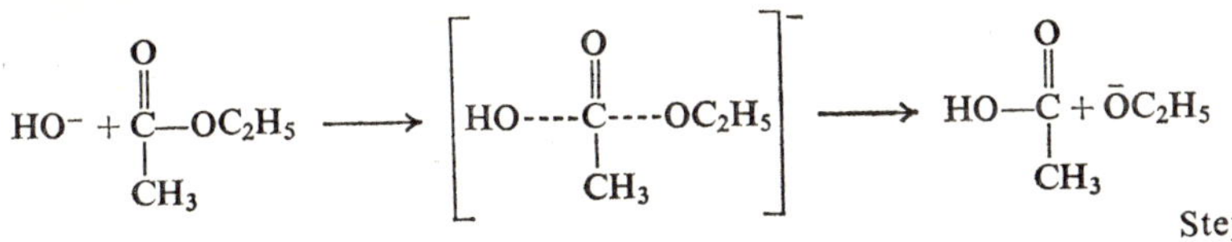

Step 1

Two further reactions are thus required to produce the observed products and therefore the reaction is not elementary if it follows this mechanism.

$$CH_3COOH + OH^- \rightarrow CH_3COO^- + H_2O \qquad \text{Step 2}$$

$$C_2H_5O^- + H_2O \rightarrow C_2H_5OH + OH^- \qquad \text{Step 3}$$

Because the observed kinetics require step 1 to be rate determining it must be postulated that steps 2 and 3 are more rapid than step 1. This is very likely since acid-base reactions of the type which occur in steps 2 and 3 are known to be very rapid. The addition of the three steps produces the stoichiometric equation for the reaction.

Since the observed rate expression cannot distinguish between these two mechanisms it is necessary to look for other evidence. The problem is to decide whether the ester is cleaved at bond 1 or bond 2 (see below). It has been found that if the *aqueous* hydrolysis of the ester is carried out in water enriched in the isotope ^{18}O, the resulting *acid* is enriched in ^{18}O and the alcohol contains only the normal isotope, ^{16}O.† This suggests that bond 1 is broken during this hydrolysis and, therefore, probably during the *alkaline* hydrolysis as well.

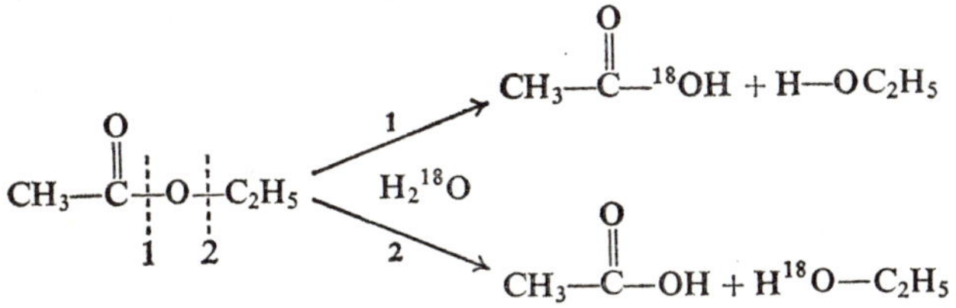

Mechanism 2 is therefore more likely than mechanism 1. This is an example of how the use of isotopes can be very important in the investigation of difficult mechanistic problems.

The hydrolysis of methyl acetate in the presence of hydrogen ions is reversible, so that the measured rate of reaction is the difference between the rate of hydrolysis and the rate of esterification. However, it is found that the esterification reaction becomes important only in the later stages of the reaction. This may be due to the high concentration of water which makes the rate of the hydrolysis very much higher than the rate of esterification for most of the reaction. Near the end of the observable reaction, equilibrium is approached as the rate of hydrolysis decreases due to the very low concentration of the remaining ester. The esterification reaction can be neglected in experiments such as 9.6.

† The possibility that the acid or alcohol, after formation, can exchange oxygen with $H_2{}^{18}O$, has been discounted experimentally.

The rate expression for the acid hydrolysis of methyl acetate was found to be

$$-\frac{d[CH_3COOCH_3]}{dt} = k_2[CH_3COOCH_3][H^+] \qquad (20)$$

but in any one experiment, the reaction was observed to be first order, with the first order rate constant equal to $k_2[H^+]$.

The hydrolysis involves water, but because its concentration in the reaction mixture is so large, the variation of the rate of reaction with changes of water concentration cannot be detected. Another possible rate expression is thus

$$-\frac{d[CH_3COOCH_3]}{dt} = k_3[CH_3COOCH_3][H^+][H_2O] \qquad (21)$$

where $k_2 = k_3[H_2O]$ [see equation (20)].

Because it is not possible to distinguish between equations (20) and (21), several mechanisms are possible.

If it could be shown that (20) is the correct rate expression, a suitable mechanism would be

Mechanism 1

$CH_3COOCH_3 + H^+ \rightarrow$ intermediate 1 — Step 1 (slow)

intermediate 1 $+ H_2O \rightarrow CH_3COOH + CH_3OH + H^+$ — Step 2 (rapid)

The rate of reaction would be equal to the rate of formation of the intermediate and this is proportional to $[CH_3COOCH_3][H^+]$, as is required by equation (20). The slow reaction in step 1 of the mechanism does not reach equilibrium because of the rapid second step.

If it could be shown that equation (21) is the correct rate expression, the following two mechanisms would provide the correct form of the rate expression.

Mechanism 2

$CH_3COOCH_3 + H^+ \leftrightharpoons$ intermediate 2 — Step 1 (rapid equilibrium)

intermediate 2 $+ H_2O \xrightarrow{k_2} CH_3COOH + CH_3OH + H^+$ — Step 2 (slow)

Step 1 is so very much more rapid than step 2 that the concentration of intermediate 2 reaches and maintains its equilibrium value throughout the reaction (except at the beginning of the reaction, when its concentration is rapidly increasing to its equilibrium value). If K is the equilibrium constant for step 1,

$$K = \frac{\text{intermediate 2}}{[CH_3COOCH_3][H^+]}$$

so that [intermediate 2] $= K[CH_3COOCH_3][H^+]$.

The rate of formation of the products is equal to the rate of the second, slow step.

$$+\frac{d[\text{products}]}{dt} = k_2[\text{intermediate 2}][H_2O]$$

$$= k_2 K[CH_3COOCH_3][H^+][H_2O]$$

which has the same form as equation (21).†

† This is a simplified treatment. (See Appendix F).

In a similar way it can be shown that mechanism 3 produces a rate expression similar to equation (21).

Mechanism 3

$$CH_3COOCH_3 + H_2O \rightleftharpoons \text{intermediate 3} \qquad \text{Step 1 (rapid equilibrium)}$$

$$\text{intermediate 3} + H^+ \rightarrow CH_3COOH + CH_3OH + H^+ \qquad \text{Step 2 (slow)}$$

In all three mechanisms, addition of the steps produces the stoichiometric equation.

It is not possible to decide between these mechanisms using only the evidence given above. However a clue may be gained from other reactions which exhibit catalysis by hydrogen ions and which involve molecules containing a carbonyl group (e.g. the iodination of acetone). It is thought that the first step in these reactions is usually

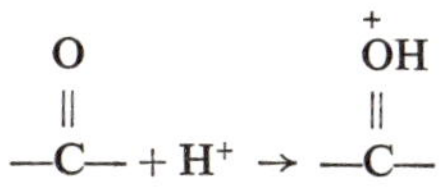

so on this basis either mechanism 1 or 2 is more likely than mechanism 3.

These mechanisms postulate the same steps, but different slow steps, so the distinction between them is unimportant to an organic chemist but poses a difficult problem for the kineticist.

Questions

1. The reaction between bromoethane and water is slow and reversible. Explain the following observations.

(*a*) The reaction between bromoethane and water in the presence of an excess of either silver nitrate or silver oxide is *slow* but *irreversible.*

(*b*) The reaction between bromoethane and water in the presence of calcium carbonate is *slow* but *irreversible.*

(*c*) The reaction between ethanol and hydrogen bromide is accelerated by the addition of concentrated sulphuric acid and is at the same time rendered irreversible.

(*d*) The reaction between bromoethane and water in the presence of lithium hydroxide is *fast* and *irreversible.*

2. The alkaline hydrolysis of iodoethane is slowed down considerably by diluting the reaction mixture with water. Why is this less effective in the alkaline hydrolysis of 2-chloro-2-methylpropane?

3. From the information given in Experiment 9.1, calculate the proportionality constant c and the second order rate constant.

4. 2-chloro-2-methylbutane is shaken with aqueous sodium nitrite solution. Suggest possible products of the reaction.

10 More aspects of the kinetics of gaseous reactions

This chapter considers some aspects of the kinetics of gaseous reactions not covered in Chapters 6 and 7. The experimental work is unfortunately too advanced for this book and it will be necessary to quote the relevant results without being able to demonstrate any of them.

The kinetics of the following two reactions will be discussed first.

Reaction	*Order with respect to reactant*
$Br_{2(g)} = 2Br_{(g)}$	2 over a wide range of pressures
cyclopropane = propene	1 when the pressure is greater than about 760 mm Hg
	2 when the pressure is less than about 760 mm Hg

If these reactions were elementary and hence unimolecular, the orders would be one under all conditions. The observed kinetics indicate therefore that the reactions are complex.

The dissociation of bromine

The order of the reaction is two so that a possible first and rate determining step in the mechanism is:

$$Br_2 + Br_2 \rightarrow ? \qquad \text{Step } 1(a) \text{ slow}$$

Experiments have shown that the rate of the dissociation is affected by the presence of other gases (M) which do not react with either bromine molecules or bromine atoms. The complete rate expression for the reaction is then:

$$-\frac{d[Br_2]}{dt} = k_2[Br_2]^2 + k_2'[Br_2][M] \qquad (1)$$

The first term in (1) is accounted for by step 1(*a*), and the second term can be taken as indicating that an alternative first and rate determining step in the mechanism could be:

$$Br_2 + M \rightarrow ? \qquad \text{Step } 1(b) \text{ slow}$$

Step 1(*a*) is now seen as the particular case of step 1(*b*) when M represents Br_2.

Since there are no new products in these steps, the role of M may be to provide sufficient energy by collision for the dissociation of bromine molecules. The first step is thus the collisional activation of the bromine molecules and the second step could be the spontaneous dissociation of the activated bromine molecules.

$$Br_2 + M \rightarrow Br_2^* + M \qquad \text{Step 1}$$

$$Br_2^* \rightarrow Br + Br \qquad \text{Step 2}$$

An asterisk denotes an activated molecule, i.e. one which possesses sufficient energy for reaction. In order to obtain the second order rate expression, step 2 must be postulated to be faster than step 1.

The isomerization of cyclopropane

If this reaction were second order under all conditions, a mechanism similar to that given for the decomposition of bromine molecules would explain the observed rate expression.

Step 1 (slow)

(M = any molecule in the reaction mixture which has sufficient energy to activate a cyclopropane molecule.)

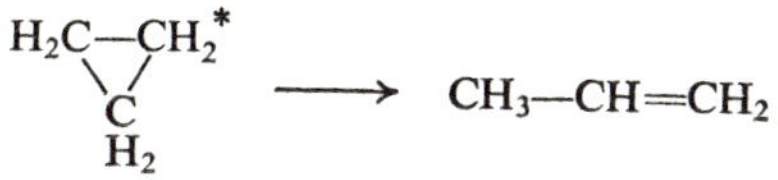

Step 2 (fast)

However, at pressures greater than about 760 mm Hg, the reaction becomes *first* order. Can the above mechanism be modified to account for this change of order? One possible modification is to assume that step 2 becomes much slower than step 1. In this situation step 1 can achieve equilibrium, the fast reverse reaction being the deactivation, by collision, of activated molecules. A slow 'leakage' of activated cyclopropane molecules occurs through step 2 (see Fig. 10.1).

The rate of formation of propene is now the rate of step 2,

i.e. $+\frac{d[\text{propene}]}{dt} = k_2$ 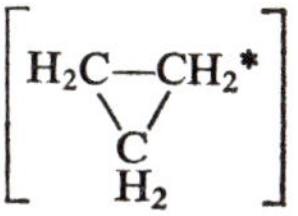 (2)

where k_2 is the rate constant for step 2. For step 1, the concentration of activated cyclopropane molecules can be expressed in terms of the equilibrium constant, K.†

$K =$ 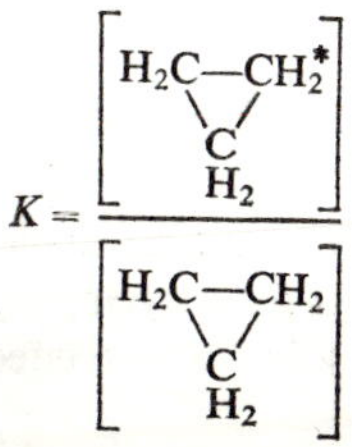 (3)

† This is a simplified treatment. (See Appendix F.)

Therefore, from (2) and (3),

$$+\frac{d[\text{propene}]}{dt} = k_2K \left[\quad \right] \qquad (4)$$

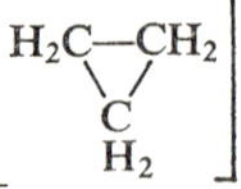

This shows that the assumption of a change in the relative rates of the two steps does lead to a first order rate expression.

However, the change from first order to second order kinetics for this reaction is dependent on pressure. Can the postulated change in the relative rates of the steps be linked with this? Moreover, for all pressures under which the reaction has been investigated, the decomposition of bromine molecules is always second order: no change of order has been detected here. Any theory must account for both types of behaviour.

Collisions and vibrations

In the mechanism represented by Fig. 10.1, step 1 is bimolecular and step 2 unimolecular. Step 1 involves the collision of two molecules with an effective kinetic energy sufficient to activate one of the molecules. In step 2 this molecule

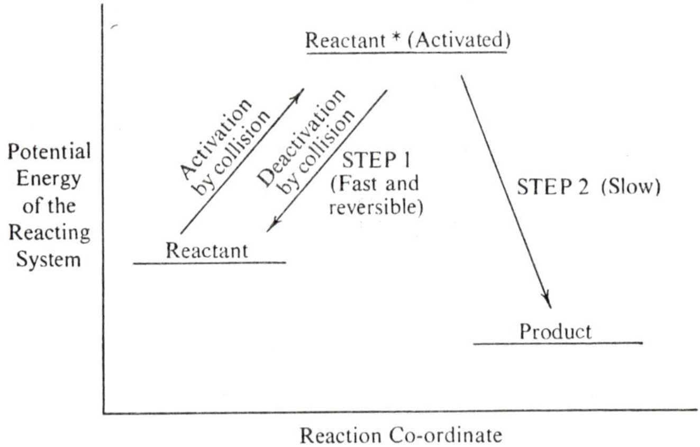

Fig. 10.1. The mechanism for the first order isomerization of cyclopropane

reacts by the breaking of one bond (as in the case of a bromine molecule) or by the rearrangement of the bonds within the molecule (as for a cyclopropane molecule). The molecules become activated by the conversion of the effective kinetic energy of collision to vibrational energy of the bonds.

Consider the simpler molecule first. The bromine atoms in a non-activated bromine molecule are thought to be rapidly vibrating. The absorption of sufficient vibrational energy to break the bond means that the next time the bromine atoms move away from each other the molecule will dissociate. It has been shown that the vibrational frequency of most bonds lies between about 10^{12} s^{-1} and 10^{14} s^{-1}. In other words, after activation has occurred the time taken for the bond to break cannot exceed about 10^{-12} s. Step 2 is thus a fast process for a bromine molecule.

How does the rate of step 2 in the decomposition of bromine compare with the rate of step 1? The collisional frequency for a bimolecular step varies with concentration and hence pressure. It can be shown that the collision frequency for one molecule is about 10^9 s^{-1} at a pressure of 760 mm Hg and a temperature of 500°C. Thus the mean time between collisions for this molecule is about 10^{-9} s and this will not vary very much with temperature. It is likely, therefore, that the activated bromine molecule will dissociate into atoms before it can be deactivated by a bimolecular collision. Or, to put this in another way, step 1 is likely to be slower than step 2 and therefore rate determining. Only if the pressure were extremely high would this theory suggest that the collision frequency would exceed the vibrational frequency. So far, the reaction has not been studied under such a pressure.

Secondly, consider the more complex cyclopropane molecule. Unlike the bromine molecule which possesses only one bond, this molecule possesses nine bonds capable of acquiring additional vibrational energy. When the molecule is activated by collision the vibrational energy may not be immediately localized in the bond which is to be broken. The molecule, on average, may have to undergo many vibrational motions before the energy is redistributed and localized in the right bonds. This takes time, and under certain conditions the activated molecule is more likely to undergo deactivating collisions than to react to give propene.

For the isomerization of cyclopropane, first order kinetics are observed at pressures above about 760 mm Hg. At this pressure the mean time between collisions is about 10^{-9} s. It must be postulated that the mean lifetime of the activated molecule is greater than 10^{-9} s so that step 2 is slower than step 1 (see Fig. 10.1).

At lower pressures, second order kinetics are observed. The frequency of collisions decreases with pressure (i.e. concentration) so that the mean time between collisions lengthens and becomes greater than the mean lifetime of the activated molecule. In this case, step 1 becomes rate determining and second order kinetics are observed.

Consider the nature of step 2 in more detail. The activated molecules are inherently unstable and can achieve a new structure without interacting with any other molecules. There is a certain probability that such a molecule will undergo the rearrangement of bonds in a given time, this probability being a function of the structure of the molecule only. This is similar to the situation in radioactive decay (see page 32) and provides an example of a step in a chemical change which is *unimolecular*: the activation (step 1) is bimolecular and the rearrangement of the activated molecule (step 2) is unimolecular.

The above theory, which was first developed by Lindemann and Hinshelwood in the 1920s, successfully explains the observed kinetics of the decomposition of bromine and the isomerization of cyclopropane (and of some other reactions) in a qualitative way. It has failed, however, to predict the magnitude of the pressure at which the change of order is observed, and precisely how this depends on the structure of the decomposing molecule.

First order reactions

The above mechanism does not necessarily apply to all gaseous reactions which exhibit first order kinetics. The decomposition of cyclobutane to ethylene is first order over a wide range of pressures and becomes second order only at very low pressures. It is thought that the mechanism of this reaction is similar to that for the isomerization of cyclopropane. However, the decomposition of dinitrogen pentoxide to nitrogen dioxide and oxygen is first order under all pressures and its stoichiometric equation involves *two* reactant molecules: the Lindemann theory applies only to those reactions whose stoichiometric equation is A = product(s) and which show first order kinetics under some conditions. Another type of mechanism has been postulated for the decomposition of dinitrogen pentoxide. The decomposition of ethane to ethylene and hydrogen is first order and shows no second order kinetics at low pressures. Various features of this reaction (see Question 3) indicate that it is probably a chain reaction. This is an example of an apparently straightforward reaction which is, in fact, very complex.

Termolecular reactions

The calculation of the frequency of three-body collisions is difficult and only an approximate value can be obtained. Some results indicate that this frequency must be at least 10^{-3} times the two-body collision frequency at about atmospheric pressure. Termolecular steps are therefore far less likely to contribute to a reaction than bimolecular steps.

Are any *elementary* termolecular reactions known? Two examples are quoted below which may be both third order and elementary, but in one case it is also possible to assume a complex reaction involving several bimolecular steps which is also in line with the observed order of the reaction.

The reaction of nitrogen monoxide and oxygen

$$2NO_{(g)} + O_{2(g)} = 2NO_{2(g)} \qquad (5)$$

The experimental rate expression is:

$$-\frac{d[O_2]}{dt} = k_3[NO]^2[O_2] \qquad (6)$$

Equation (5) could be explained by a single termolecular step. However, another mechanism consistent with the rate expression is:

$$2NO \rightleftharpoons (NO)_2 \qquad \text{Step 1 (fast in both directions)}$$

$$(NO)_2 + O_2 \rightarrow 2NO_2 \qquad \text{Step 2 (slow)}$$

This mechanism gives a rate expression which can be shown to have the form of equation (6).†

† See Appendix F.

The complex mechanism is reasonable because a nitrogen monoxide molecule possesses one unpaired electron, and two molecules could therefore associate to form a dimer.

The association of bromine atoms

$$2Br_{(g)} = Br_{2(g)} \tag{7}$$

The rate expression for this reaction in the presence of an unreactive foreign gas (M, e.g. Ar) has been found for a wide range of temperatures to be:

$$-\frac{1}{2}\frac{d[Br]}{dt} = k_3''[Br]^2[M] + k_3[Br]^2[Br_2] \tag{8}$$

In the absence of M the rate expression does not contain the first term. This result is surprising since it might have been expected that the reaction would consist of the combination of two bromine atoms thus giving a *second* order rate expression.

Equation (8) can be interpreted by assuming the reaction occurs via two separate competing pathways (see page 78).

$$Br + Br + M \rightarrow Br_2 + M \tag{9}$$

$$Br + Br + Br_2 \rightarrow Br_2 + Br_2 \tag{10}$$

Both reactions are elementary and termolecular. The fact that bromine atoms cannot combine except in the presence of a third body is known as the *third-body restriction.* For a diatomic molecule a bimolecular association of its atoms is not possible because the energy liberated by the association cannot be stored in the molecule. The energy is necessarily absorbed in the Br—Br bond and it will always be sufficient to cause immediate dissociation. A third body is required to absorb the energy released as kinetic energy of translation. The third-body restriction does not apply to larger molecules which are capable of redistributing the vibrational energy throughout several bonds.

Chain reactions

These are reactions showing certain features which are most easily explained by a chain mechanism (see below). No one chain reaction will necessarily show all the distinguishing features listed below and some reactions not involving a chain mechanism may occasionally show some of the following features.

Characteristics of chain reactions

1. The rates of chain reactions are markedly affected by irradiation of the reaction mixture using light of a certain wavelength which can be absorbed by one of the reactants, so producing dissociation. The *quantum yield* of such a reaction is the number of molecules of reactant decomposed for each quantum of light absorbed. This can be as high as 10^6 for the hydrogen/chlorine reaction, and many other chain reactions have high quantum yields.
2. In some chain reactions there is an *induction period* before any measurable change in pressure or concentration occurs. The rate may then increase rapidly

or slowly depending on the type of reaction (see Fig. 10.2). A reaction which does not proceed by a chain mechanism usually has a rate which decreases as the reactant concentrations decrease. (Autocatalytic reactions are an exception.)

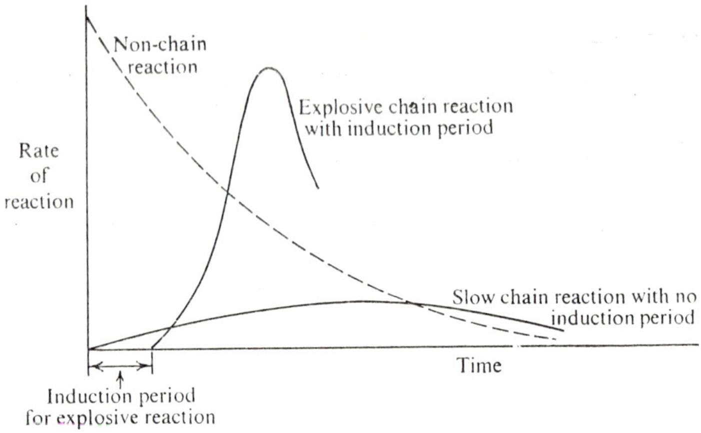

Fig. 10.2

3. Traces of other substances can greatly affect the rate of a chain reaction. This fact makes the experimental study of chain reactions very difficult. Even when very pure reactants are used, the rate of reaction may still be affected by the material from which the reaction vessel was made, its shape and even its history. Under certain conditions, for instance, the rate of reaction between hydrogen and chlorine is proportional to the square of the diameter of a cylindrical reaction vessel. The wall of the vessel also acts as an unremovable inhibitor in certain reactions. Initiators and inhibitors have already been discussed in Chapter 8.
4. Chain reactions usually show complex kinetics. It is common to find non-integral orders and complicated rate expressions in which a simple total order cannot be given. On the other hand, a few do show simple first order kinetics. The rate expressions for three chain reactions are given below.

$$C_2H_{6(g)} = C_2H_{4(g)} + H_{2(g)} \qquad -\frac{d[C_2H_6]}{dt} = k_1[C_2H_6]$$

$$CH_3CHO_{(g)} = CH_{4(g)} + CO_{(g)} \qquad -\frac{d[CH_3CHO]}{dt} = k[CH_3CHO]^{3/2}$$

$H_{2(g)} + Br_{2(g)} = 2HBr_{(g)}$ See Table 6.1 (page 70)

5. Free radicals and atoms have been detected in some chain reactions. One method for detecting alkyl radicals is to react them with a metallic 'mirror' to form a volatile metal alkyl, e.g.

$$Zn + 2CH_3 = Zn(CH_3)_2$$

Chain mechanisms

The reaction between hydrogen and chlorine was one of the first chain reactions to be studied in detail. Nernst found that this reaction possessed many of the above features and proposed in 1913 that a *chain mechanism* could explain the observations. A later and very detailed study of the kinetics of the hydrogen–bromine reaction again indicated that a chain mechanism was possible. The rate expression is extremely complicated (see Table 6.1). Both mechanisms are thought to be essentially similar but the hydrogen–chlorine reaction is much faster and is complicated by the presence of traces of moisture and oxygen as well as being very sensitive to light.

The first step (the *initiation*) is postulated to be the dissociation of the halogen molecule.

$Br_2 \rightarrow Br + Br$ initiation†

The energy required for the Br—Br bond to be broken comes from thermal energy (the reaction is then called the thermal reaction), or from irradiation by light (the photochemical reaction). The highly active free atoms may then react with hydrogen molecules.

$Br + H_2 \rightarrow HBr + H$ propagation

and then

$H + Br_2 \rightarrow HBr + Br$ propagation

These two propagation steps can therefore be repeated many times because both hydrogen and bromine atoms are regenerated. These types of mechanism are called *chain* mechanisms because of the nature of these two steps. The hydrogen and bromine atoms are known as *chain centres*. The two propagation steps continue until one or other centre is removed in a *termination* step such as

$2Br = Br_2$ termination‡

It is clear that other steps are also possible such as:

$H_2 \rightarrow H + H$ initiation

$H + HBr \rightarrow H_2 + Br$

$H + Br \rightarrow HBr$ termination‡ etc.

All chain reactions are considered to consist of three stages viz. initiation, propagation, and termination. In the propagation steps, chain centres are continually regenerated. Non-chain mechanisms do involve reactive intermediate species, but these are not regenerated so that propagation is not possible. The high rates of many chain reactions can be ascribed to these propagation steps. When a chain reaction is relatively slow it is assumed either that the initiation is low or that the chain centres are removed very quickly after their formation, possibly by an inhibitor.

† In all reactions between hydrogen and the halogens the dissociation of hydrogen molecules is not so likely as the dissociation of halogen molecules because of the high dissociation energy of 432 kJ mol^{-1} for hydrogen.

‡ This should be written with a 'third body'.

A detailed mechanism for the hydrogen–bromine reaction

This can only be postulated after considering the experimental rate expression and the 'likelihood' of a particular step occurring. Table 6.1 gives details of the rate expression for this reaction as well as for the hydrogen–chlorine and hydrogen–iodine reactions.

Both the thermal and the photochemical reactions show that the *initial* rate is proportional to $[H_2][Br_2]^{1/2}$. As the reaction proceeds the rate falls off more rapidly than would be the case if the initial rate expression held throughout the reaction. When hydrogen bromide is added to the initial reaction mixture the initial rate is lowered. The reversibility of this reaction might seem to account for these facts, but at normal temperatures the equilibrium is well over to the product side, and the reverse reaction is negligible for most of the reaction. A better interpretation is that the mechanism involves a step in which some of the hydrogen bromide reacts to regenerate one of the reactants so opposing the main trend of the reaction, e.g.

$$H + HBr \rightarrow H_2 + Br \qquad E \sim 16\ \text{kJ mol}^{-1}$$

(*E* is the activation energy in each case)

or

$$Br + HBr \rightarrow Br_2 + H \qquad E \sim 176\ \text{kJ mol}^{-1}$$

The first reaction has a much lower activation energy and will be much faster. It is therefore more likely to make a significant contribution to the mechanism.

A possible complete mechanism for the reaction is:

$Br_2 \rightarrow 2Br$	initiation 1
$Br + H_2 \rightarrow HBr + H$	propagation 2
$H + Br_2 \rightarrow HBr + Br$	propagation 3
$H + HBr \rightarrow H_2 + Br$	inhibition 4 (reverse of 2)
$2Br \rightarrow Br_2$	termination† 5

The justification for this mechanism is that the observed rate expression can be obtained from it.

All mechanisms discussed in previous chapters have been treated mathematically so as to yield a rate expression for the reaction. The assumption made was that one step is slower than the others and therefore determines the rate. In a chain reaction no such assumption can be made. The relative importance of each step varies throughout the reaction. At first the rate is low because the rate of initiation is low, but as the chain centres are produced the reaction accelerates and the rate ceases to depend only on the rate of initiation. A further consequence of this is that a chain mechanism does not have individual steps which can be summed to give the stoichiometric equation.

When no step is assumed to be rate determining throughout the reaction another simplifying assumption may be made so that a rate expression can be obtained from the mechanism. This is the *steady (or stationary) state approximation,* which assumes that the intermediates are so reactive that their concentrations

† A 'third body' is required in this reaction.

remain very low and effectively constant compared with the reactant concentrations. The change in the concentrations of the intermediates with time is therefore assumed to be zero. For example, in the above mechanism the hydrogen atom and the bromine atom concentrations are assumed constant so that:

$$-\frac{d[\mathrm{H}]}{dt}=0 \quad \text{and} \quad -\frac{d[\mathrm{Br}]}{dt}=0$$

This assumption is applicable to chain reactions because the intermediates are highly reactive free radicals or atoms.

From the above mechanism,

$$+\frac{d[\mathrm{HBr}]}{dt}=\mathrm{Const.}[\mathrm{Br}][\mathrm{H_2}]+\mathrm{Const.}[\mathrm{H}][\mathrm{Br_2}]-\mathrm{Const.}[\mathrm{H}][\mathrm{HBr}] \qquad (11)$$

Using the steady state approximation, expressions for [H] and [Br] can be derived so that equation (11) can be written in terms of $[H_2]$, $[Br_2]$ and [HBr]. When this is done the expression obtained is of the same form as the experimental rate expression showing that the mechanism is consistent with the observed kinetics.

A comparison of the reactions between hydrogen and the halogens

Why is it that the hydrogen–chlorine reaction is a fast, often explosive, chain reaction, the hydrogen–bromine reaction a slower chain reaction, and the hydrogen–iodine reaction a non-chain reaction? Such diversity of kinetic behaviour amongst similar chemical reactions needs explaining. Some relevant data are given in Table 10.1.

Table 10.1. The hydrogen–halogen reactions. Some approximate activation energies for elementary steps

	H_2—Cl_2	H_2—Br_2	H_2—I_2
Experimental E (kJ mol^{-1}) for $H_2 + X_2 = 2HX$ the overall reaction	209	171	167
Energy required to dissociate a diatomic molecule (kJ mol^{-1}) $X_2 \rightarrow X + X$ (i.e. for initiation step)	242	188	146
Experimental E (kJ mol^{-1}) for first propagation step $H_2 + X \rightarrow HX + H$	25	75	142

Less energy is required to dissociate an iodine molecule than a bromine or chlorine molecule so it should be easiest to initiate a hydrogen–iodine chain reaction. However, the essential feature of a spontaneous chain reaction is a series of fast propagation steps. If the propagation steps are potentially slow then an ordinary molecular (i.e. non-chain) mechanism may be preferred. The first propagation step for the hydrogen–chlorine reaction has a low activation energy.

It is therefore likely to be very fast. On the other hand, the reaction $H_2 + I \rightarrow HI + H$ has a fairly high activation energy, so it is likely to be slow. In fact, the experimental activation energy (E) of this step is almost as high as the activation energy for the complete reaction. It is likely, therefore, that the hydrogen–chlorine reaction will proceed by a chain mechanism and the hydrogen–iodine reaction by a non-chain mechanism. The hydrogen–bromine reaction is seen to fall somewhere between the hydrogen–chlorine and hydrogen–iodine reactions. Its first propagation step has a much lower activation energy than that for complete reaction. The reaction is a chain reaction but it is slower than the hydrogen–chlorine reaction.

The hydrogen–iodine reaction was thought for a long time to be an elementary, bimolecular reaction in view of its rate expression (see Table 6.1). However, recent work† has shown that the first step probably involves the dissociation of iodine molecules, so that a possible mechanism is:

$I_2 \rightleftharpoons 2I$ Step 1 (fast in both directions)

$2I + H_2 \rightarrow 2HI$ Step 2 (slow)

Note that this is not a chain reaction. Although iodine atoms are involved they are not regenerated in a propagation step. The rate of production of hydrogen iodide is given by:

$$+\frac{1}{2}\frac{d[HI]}{dt} = \text{Const.}[H_2][I]^2$$

Since from step 1, K, the equilibrium constant, $= [I]^2/[I_2]$,

$$+\frac{1}{2}\frac{d[HI]}{dt} = \text{Const.}[H_2][I_2]$$

The mechanism is therefore consistent with the observed kinetics.

Branched chain reactions

In these reactions, the reaction of *one* chain centre in a propagation step produces *two* (or more) new chain centres instead of the usual *one*. For instance, in the combustion of hydrogen and hydrocarbons, possible steps in the mechanism are:

$\underline{O} + H_2 \rightarrow \underline{H} + \underline{OH}$

$\underline{H} + O_2 \rightarrow \underline{O} + \underline{OH}$

$\underline{OH} + H_2 \rightarrow \underline{H} + H_2O$ (chain centres underlined)

As the reaction proceeds, more and more chain centres are formed, and the reaction is accelerated unless the termination steps can remove the centres at least as quickly as they are formed. The rate may increase so rapidly that the reaction becomes uncontrollably fast. Since most chain reactions are exothermic, the rate will also be increased by the rise in temperature, so that the resulting explosion is due both to the branched chain mechanism and the heat produced.

† *J. Chem. Phys.*, 1967, **46**, 77.

Fire extinguishers

Fire extinguishers frequently work either by excluding oxygen or by lowering the temperature of the combustible material. However, it is found that very low concentrations (of the order of 5 per cent by volume) of some halogenated paraffins are effective in stopping combustion. Examples of these are CCl_4, CH_3Br, and $CClBrF_2$('B.C.F.').

How do these halogenated paraffins work? It is thought that they undergo decomposition to give products which react with atoms and radicals involved in the chain reaction: they thus prevent propagation. The most important inhibitors may be the hydrogen halides which can remove OH radicals by the reaction:

$$OH + HX \rightarrow H_2O + X \qquad (X = \text{a halogen atom})$$

The halogen atoms can then reform HX molecules so regenerating the inhibitor.

Questions

1. 'Chain reactions can be considered to be reactions which generate their own catalyst.' Discuss this statement, indicating the possible catalyst(s) in the hydrogen–bromine mechanism. In what way does this kind of 'autocatalysis' differ from the usual type of autocatalysis discussed in Chapter 8?

2. The decomposition of acetaldehyde vapour

$$CH_3CHO_{(g)} = CH_{4(g)} + CO_{(g)}$$

gives the following rate expression:

$$-\frac{d[CH_3CHO]}{dt} = \text{Const.}[CH_3CHO]^{3/2}$$

A proposed mechanism is:

$$CH_3CHO \xrightarrow{k_1} CH_3^{\bullet} + {}^{\bullet}CHO \qquad \text{Step 1}$$

$$CH_3CHO + CH_3^{\bullet} \xrightarrow{k_2} CH_4 + CO + CH_3^{\bullet} \qquad \text{Step 2}$$

$$CH_3^{\bullet} + CH_3^{\bullet} \xrightarrow{k_3} C_2H_6 \qquad \text{Step 3}$$

Is this an example of a chain mechanism? Using the steady state approximation for CH_3

$$\left(-\frac{d[CH_3]}{dt} = 0\right)$$

and the fact that the rate of reaction is given by

$$+\frac{d[CH_4]}{dt} = k_2[CH_3][CH_3CHO]$$

(from step 2) show that the mechanism is in agreement with the observed rate expression.

3. Using the following evidence discuss the possibility of the decomposition of ethane being a chain reaction.

$$C_2H_{6(g)} = C_2H_{4(g)} + H_{2(g)}$$

Rate expression:

$$-\frac{d[C_2H_6]}{dt} = k_1[C_2H_6]$$

The rate is markedly accelerated by traces of ether and slowed down by traces of nitrogen monoxide. The reaction is unaffected by ultra-violet light with a wave-length greater than 250 nm (2500 Å). If you think it is a chain reaction suggest a likely initiation step.

11 Kinetics and equilibrium

Consider the reversible reaction with the stoichiometry

$$aA + bB \leftrightharpoons cC + dD \quad (1)$$

which is started by mixing the reagents A and B. The rate of the forward reaction (A reacting with B) is at a maximum initially but decreases as the concentrations of A and B decrease. The rate of the reverse reaction (C reacting with D) is initially zero but increases as the concentrations of C and D increase from zero. If sufficient time is allowed, the rates of the forward and reverse reactions will become equal and the composition of the reaction mixture will then remain constant, provided the temperature is kept constant. The system is then said to be in a state of dynamic equilibrium.

The equilibrium constant, K, of the reaction at that temperature is given by

$$K = \frac{[C]_e^c[D]_e^d}{[A]_e^a[B]_e^b} \; \dagger \quad (2)$$

where the subscript e denotes the concentration at equilibrium. The form of equation (2) has been verified experimentally for many reactions and has also been derived theoretically from thermodynamics.

The question now arises as to whether there is any connection between the equilibrium constant expression and the rate expressions for the forward and reverse reactions. The following reaction, the oxidation of iodide ions with arsenic acid in aqueous solution, will be discussed in this connection.

$$H_3AsO_{4(aq)} + 2I^-_{(aq)} + 2H^+_{(aq)} \rightleftharpoons H_3AsO_{3(aq)} + I_{2(aq)} + H_2O_{(l)}$$

The observed rate expressions are:

$$\text{Rate of forward reaction} = k_f[H_3AsO_4][I^-][H^+]$$

$$\text{Rate of reverse reaction} = k_r\frac{[H_3AsO_3][I_2]}{[I^-][H^+]}$$

where k_f and k_r are the rate constants for the forward and reverse reactions. At equilibrium, the rates are equal, so that

$$k_f[H_3AsO_4][I^-][H^+] = k_r\frac{[H_3AsO_3][I_2]}{[I^-][H^+]}$$

and

$$\frac{k_f}{k_r} = \frac{[H_3AsO_3][I_2]}{[H_3AsO_4][I^-]^2[H^+]^2} \quad (3)$$

The equilibrium constant expression for the reaction is:

$$K' = \frac{[H_3AsO_3][I_2][H_2O]}{[H_3AsO_4][I^-]^2[H^+]^2} \quad (4)$$

† The quantities used for the evaluation of K should be the activities of the species rather than their molar concentrations.

Equations (3) and (4) are identical except for the factor $[H_2O]$ in (4). The rate of the reverse reaction may be proportional to the concentration of water, but this is impossible to detect because of the very high, and hence effectively constant, concentration of water in the experiments. For the same reason, the equilibrium expression for a dilute solution of the reaction mixture may be written

$$K = \frac{[H_3AsO_3][I_2]}{[H_3AsO_4][I^-]^2[H^+]^2},$$

where $K = K'/[H_2O]$.

Thus a qualitative connection between the rate expressions and the equilibrium constant expression is established. The following results test the relationship quantitatively:

K at 0°C = 6·2 $mol^{-3}\ dm^9$

k_f at 0°C = $1{\cdot}05 \times 10^{-4}\ mol^{-2}\ dm^6\ s^{-1}$

k_r at 0°C = $1{\cdot}57 \times 10^{-5}\ mol\ dm^{-3}\ s^{-1}$

Thus $k_f/k_r = 6{\cdot}7\ mol^{-3}\ dm^9$, which is within 10 per cent of the observed value for K.

Another reversible reaction which has been thoroughly investigated is that between hydrogen and iodine.

$$H_{2(g)} + I_{2(g)} \rightleftharpoons 2HI_{(g)}$$

The form of the equilibrium expression can be obtained by equating the observed forward and reverse rate expressions.

$$k_f[H_2][I_2] = k_r[HI]^2$$

and

$$\frac{k_f}{k_r} = \frac{[HI]^2}{[H_2][I_2]} = K$$

Bodenstein found a value of about 48 for the equilibrium constant at 716°C, and the ratio k_f/k_r at the same temperature is accurately 40·2. These values differ by about 15 per cent.

Unfortunately, few reversible reactions have been investigated with the aim of comparing the ratio of the rate constants for the forward and reverse reactions with the equilibrium constant. The examples which have been tested in this way show that there is always agreement between the *forms* of the expressions for K and k_f/k_r, and the quantitative agreement is accepted as being as good as experimental error will allow (although this may not be so for the hydrogen–iodine reaction). The relationship,

$$K = \frac{k_f}{k_r} \qquad (5)$$

can be accepted as valid so long as it is clear just what k_f and k_r refer to (see page 154), and it has been used to calculate k_f for a reaction from the experimental values of k_r and K (see Question 3 for an example).

However, equation (5) must be used with caution. Measurements of the rates of reaction are almost always made with the reaction in a non-equilibrium state.† Thus the observed rate expressions apply to non-equilibrium mixtures. But because the equilibrium constant expression applies only at equilibrium, the rate expressions used to evaluate k_f/k_r to test equation (5) should be those ***which apply at equilibrium.***

In the oxidation of iodide ions by arsenic acid and the reaction between hydrogen and iodine, the implicit assumption was that the rate expressions derived from experiments on non-equilibrium mixtures, apply also at equilibrium. In these cases the assumption is justified by the results, but there are reactions for which the rate expression is known to change as the reaction proceeds.

The decomposition of nitrosyl chloride

The reaction is reversible and has been studied extensively.‡

$$2ONCl_{(g)} \leftrightharpoons 2NO_{(g)} + Cl_{2(g)}$$

The experimental rate expressions are as follows:

$$-\frac{1}{2}\frac{d[ONCl]}{dt} = k_1[ONCl]^2 \qquad \text{for the forward reaction} \qquad (6)$$

and

$$+\frac{1}{2}\frac{d[ONCl]}{dt} = k_2[NO]^2[Cl_2] \qquad \text{for the reverse reaction} \qquad (7)$$

The rate expressions are those to be expected if the reaction is elementary with the course of the reaction being pictured as:

Mechanism 1

$$O{=}N{-}Cl + Cl{-}N{=}O \rightleftharpoons O{=}N\cdots Cl\cdots Cl\cdots N=O \rightleftharpoons O{=}N + Cl{-}Cl + N{=}O$$

Broken lines represent bonds which are being made or broken in the activated complex.

At equilibrium, the rates of the two reactions are equal and equating the two rate expressions (6) and (7) shows that the equilibrium constant of the reaction K is given by $K = k_1/k_2$ [see equation (5)]. However in 1953, evidence of another possible mechanism was obtained. The addition of non-reacting gases such as nitrogen or carbon dioxide to nitrosyl chloride was found to increase its rate of decomposition. The rate expression found to apply at all stages in the forward reaction was:

$$-\frac{1}{2}\frac{d[ONCl]}{dt} = k_3[M][ONCl] \qquad (8)$$

where M represents the concentration of added gas.

† The rate of exchange of a radioactive isotope between reactants and products can be used to measure the rate of a reaction at equilibrium. Relaxation techniques measure rates near to, but not at, equilibrium.

‡ See *Chemistry Student*, 1967, **1**, 10.

These observations were explained by postulating that nitrosyl chloride molecules could be decomposed into nitrogen monoxide molecules and chlorine atoms by sufficiently vigorous collisions with other molecules† (M) in the reaction mixture. The extremely reactive chlorine atoms then rapidly react with nitrosyl chloride molecules.

Mechanism 2F (forward)

$ONCl + M \rightleftharpoons ON + Cl + M$	Step 1 (slow)	(9)
$ONCl + Cl \rightleftharpoons ON + Cl_2$	Step 2 (fast)	(10)

The overall rate of the reaction is equal to the rate of the slow step (9) and this accounts for the observed rate expression (8). Addition of the equations for the two steps gives the stoichiometric equation of the reaction.

Investigations showed that the initial rate of the reverse reaction under these conditions is:

$$+\frac{1}{2}\frac{d[ONCl]}{dt} = k_4[Cl_2][NO] \qquad (11)$$

The combination of equations (8) and (11) shows that k_3/k_4 is not equal to the equilibrium constant expression for the reaction. A possible explanation is that either one or both of equations (8) and (11) do not apply at equilibrium. Further experiments showed that after about 10 per cent of the reverse reaction had occurred, the rate expression for the reverse reaction changes to

$$+\frac{1}{2}\frac{d[ONCl]}{dt} = k_5\frac{[NO]^2[Cl_2][M]}{[ONCl]} \qquad (12)$$

There is then no further detectable change in the rate expression. Equating the rate expressions (8) and (12) shows that k_3/k_5 equals K which is in agreement with equation (5). Thus the rate expression (12), not (11), must be used to obtain equation (5), because this is the one which applies at equilibrium.

How can this change in the rate expression of the reverse reaction be explained? The mechanism of the reverse reaction is the reverse of mechanism 2F:

Mechanism 2R (reverse)

$Cl_2 + NO \rightleftharpoons Cl + ClNO$	fast	(13)
$M + Cl + NO \rightleftharpoons M + ClNO$	slow	(14)

Both reactions produce nitrosyl chloride, but (14) can occur only after (13) has produced some chlorine atoms. Since the concentration of nitrosyl chloride is initially zero, reaction (13) cannot reach an equilibrium state immediately and the observed rate of reaction is given by the rate of reaction (13) i.e. constant$[Cl_2][NO]$, which has the form of rate expression (11). As the reaction proceeds, the concentration of nitrosyl chloride increases from zero. Since reaction (13) is fast and reversible, and reaction (14) is slow, it is likely that reaction (13) will eventually reach an equilibrium state and that there will be a

† Any of the molecules present in the reaction mixture can fulfil the role of M in step 1, even other ONCl molecules, so that in the mathematical analysis of the results k_3 [M] was replaced by the sum of the several necessary terms.

slow 'leakage' of chlorine atoms through reaction (14). The rate of reaction is then the rate of the slow step (14) and is equal to

$$k_6[\mathrm{M}][\mathrm{Cl}][\mathrm{NO}]. \tag{15}$$

The equilibrium constant, K, of reaction (13) is given by

$$K = \frac{[\mathrm{Cl}][\mathrm{ClNO}]}{[\mathrm{Cl_2}][\mathrm{NO}]}$$

so that

$$[\mathrm{Cl}] = K\frac{[\mathrm{Cl_2}][\mathrm{NO}]}{[\mathrm{ClNO}]}$$

Substituting for [Cl] in (15) gives

$$+\frac{1}{2}\frac{d[\mathrm{ONCl}]}{dt} = \frac{k_6 K[\mathrm{M}][\mathrm{Cl_2}][\mathrm{NO}]^2}{[\mathrm{ClNO}]}$$

which has the form of the observed rate expression (12).†

Thus the difference between the initial rate expression and the rate expression for the reaction in its later stages is due to the increasing importance of the reaction

$$\mathrm{Cl} + \mathrm{ClNO} \rightarrow \mathrm{Cl_2} + \mathrm{NO}$$

as the concentration of nitrosyl chloride increases from zero.

It should now be seen that the equilibrium constant expression can be obtained only by combining the forward and reverse rate expressions for reactions which are the exact reverse of each other. Mechanisms 2F and 2R are not the exact reverse of each other initially so that rate expressions (8) and (11) cannot produce the equilibrium constant expression. But mechanisms 2F and 2R (after 10 per cent of the reaction has occurred), which are the exact reverse of each other, correspond to rate expressions (8) and (12) and these do give the form of the equilibrium constant expression.

The decomposition of nitrosyl chloride can therefore occur by an elementary process (mechanism 1) or by a complex process (mechanism 2), and provided the rate expressions which apply at equilibrium are used the relationship $K = k_f/k_r$ holds for each *separate* mechanism.

The composition of the equilibrium mixture can be evaluated from the results of the kinetics experiments. At 500 K, for instance, each mechanism has been found to give an equilibrium composition of 56 ± 1 per cent nitrosyl chloride. The rates of the forward and reverse reactions of mechanism 1 and of mechanism 2 become equal at the *same* composition.

This result is in accordance with a general principle known as the principle of microscopic reversibility, which states that in a mixture at equilibrium, the contribution to the rate of the forward reaction due to a certain mechanism must exactly equal the contribution of the same mechanism to the rate of the reverse

† See Appendix F for a fuller treatment.

reaction. It is impossible to have the bulk of the forward reaction at equilibrium proceeding by one mechanism, and the bulk of the reverse reaction by another mechanism. The principle indicates what has already been concluded: the relationship $K = k_f/k_r$ *must* hold so long as k_f and k_r are the rate constants which apply to *individual* mechanisms *at equilibrium.*

However, this principle does not necessarily apply in non-equilibrium mixtures. In the above example, at 500 K, only 15 per cent of the initial forward reaction occurs by mechanism 2, whereas in the equivalent mixture of nitrogen monoxide and chlorine, at the same temperature, 84 per cent of the initial reverse reaction occurs by mechanism 2. At equilibrium, the contribution of mechanism 2 to the forward rate has increased to 20 per cent, while the contribution of mechanism 2 to the reverse reaction has decreased to 20 per cent.

Entirely separate measurements of the equilibrium constant for this reaction at 500 K give an equilibrium concentration of nitrosyl chloride equal to $56{\cdot}0 \pm 0{\cdot}5$ per cent, which is in excellent agreement with the results obtained from the kinetic work.

The reasons for the rate expression for a reaction changing as the reaction proceeds can now be given.

1. A reaction may occur by two or more independent mechanisms (each with its own rate expression). One mechanism may predominate in the early stages of the reaction, whilst another becomes more important in the later stages of the reaction.
2. A reaction may occur by one complex mechanism, and the observed rate expression changes due to the changes in the concentrations of the reactants or products, as the reaction proceeds. In such an example, the contribution of one step in the mechanism to the total rate of reaction changes markedly during the reaction.

These reasons for the possible changes in the rate expressions of reactions emphasize the care that is needed in the use of equation (5). The difficulty is to identify and allow for these changes in mechanism. The easier approach is to determine a forward and a reverse rate expression in non-equilibrium mixtures and then to equate these in order to test equation (5). If equation (5) holds qualitatively and quantitatively it can be assumed that there is no mechanism change as equilibrium is approached (see the reaction between arsenic acid and iodide ions in acid solution). If it does not hold, further investigations are required.

However, it should be clear that these complications cannot apply to elementary reactions, but here the difficulty is in the identification of truly elementary reactions: many reactions at one time thought to be elementary have subsequently been shown to be complex.

The Law of Mass Action

In 1879 Guldberg and Waage put forward their Law of Mass Action. The exact meaning of their proposals has been the subject of some debate, but the treatment frequently given will be stated and then discussed here.

For a reaction with the stoichiometry

$$aA + bB \rightleftharpoons cC + dD$$

which takes place in a homogeneous system, the Law of Mass Action states that the rate expressions for the forward and reverse reactions can be predicted from the stoichiometric equation and are given respectively by

$$k_f[A]^a[B]^b \qquad \text{and} \qquad k_r[C]^c[D]^d$$

where k_f and k_r are the rate constants for the forward and reverse reactions. At equilibrium, the rates of the forward and reverse reactions are equal, so

$$k_f[A]^a[B]^b = k_r[C]^c[D]^d$$

and

$$\frac{k_f}{k_r} = \frac{[C]^c[D]^d}{[A]^a[B]^b}$$

It can be seen that the equilibrium constant, K, for the reaction is given by

$$K = \frac{k_f}{k_r}$$

which is identical with equation (5) obtained above.

This successful kinetic 'derivation' of the equilibrium constant expression has sometimes been taken as proof of the validity of the 'derivation'. However, it should be noted that there are many other pairs of rate expressions which give the correct form of K, e.g.

$$k_f[A]^a = k_r\frac{[C]^c[D]^d}{[B]^b}$$

or

$$k_f[A]^{2a}[B]^{2b} = k_r[C]^{2c}[D]^{2d}$$

Moreover, even a cursory examination of the rate expressions for several reactions shows that it is not usually possible to predict the rate expression for a reaction from its stoichiometric equation. (The examples discussed previously in this book are in no way unrepresentative.) The only reactions for which it can be done are elementary ones, but there is no *a priori* way of knowing whether a reaction is elementary or complex. There is therefore no justification for the prediction of rate expressions from the stoichiometric equation.

A further difficulty is that, even when the rate expression for a reaction is found to be 'as predicted' by the Law of Mass Action, this reaction is not necessarily elementary. The alkaline hydrolysis of esters is one example (see page 132), and the reaction between hydrogen and iodine is another (see page 146).

It cannot be emphasized too strongly that rate expressions cannot be predicted but can only be obtained from experimental rate measurements.

Questions

1. For the reaction

$$3HNO_{2(aq)} \rightleftharpoons H^+_{(aq)} + NO^-_{3(aq)} + 2NO_{(g)} + H_2O_{(l)}$$

the observed rate expressions for the forward and reverse reactions are respectively $k_f[HNO_2]^4/P_{NO}^2$ and $k_r[H^+][NO_3^-][HNO_2]$, where P_{NO} represents the partial pressure of nitric oxide above the solution. Show that these expressions are consistent with the equilibrium constant, K, being given by $K = k_f/k_r$.

2. For the reaction

$$CH_3COCH_{3(aq)} + I_{2(aq)} \rightleftharpoons CH_3COCH_2I_{(aq)} + I^-_{(aq)} + H^+_{(aq)}$$

the forward rate expression is given by $k_f[CH_3COCH_3]$ for any one run. The reverse rate expression has never been determined. *If* K equals k_f/k_r for this reaction what is the likely form of the reverse rate expression? Comment on the significance of the term $[I_2]$ in the denominator.

3. The ionic product for water, K_w, at 25°C is $1{\cdot}0 \times 10^{-14}$ mol² dm⁻⁶. Using this value, calculate the equilibrium constant, K, for the reaction

$$H_2O_{(l)} \rightleftharpoons H^+_{(aq)} + OH^-_{(aq)}$$

This reaction is elementary so $K = k_f/k_r$ and the experimental value of k_r is $1{\cdot}4 \times 10^{11}$ mol^{-1} dm^3 s^{-1}. Calculate the value of the rate constant for the forward reaction, giving its units. Why is k_r much greater than k_f?

Appendix A Additional comments on some experimental results

The reaction between metals and dilute acid (Experiments 2.2 to 2.6)

For Experiment 2.6, zinc is chosen instead of magnesium. The rate of reaction of magnesium in dilute hydrochloric acid or dilute sulphuric acid is affected by the addition of metallic salts, but at room temperature the effect is of a slight *slowing* of the reaction. At higher temperatures the rate of reaction is markedly increased by the addition of salts (see Question 3 in the exercises at the end of Chapter 2). In contrast, copper sulphate solution has a marked accelerating effect on the rate of the reaction between zinc and dilute sulphuric or hydrochloric

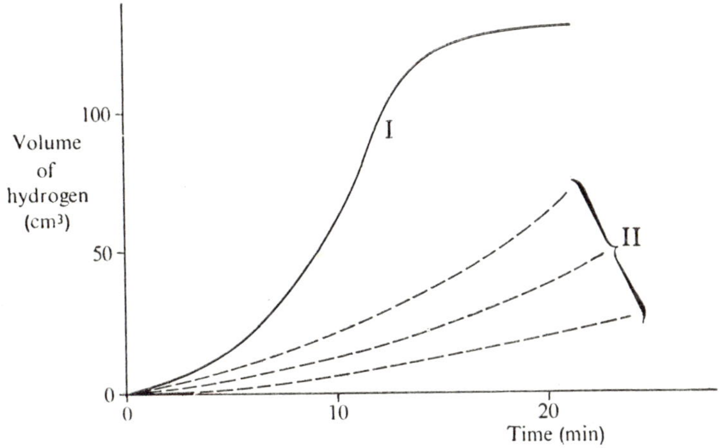

Fig. A.1. Some results for Experiment 2.6 (concentrations as in text)

I. Typical curve for Zn—HCl—$CuSO_4$
II. Possible curves for Zn—HCl only

acid at room temperature and above. This is therefore a more suitable experiment for an elementary study.

Fig. A.1 shows the wide variety of curves which can be expected for Experiment 2.6, depending on the surface area, the degree of tarnishing, and the purity of the zinc granules.

The reaction between thiosulphate and hydrogen ions (Experiments 3.2 and 3.3)

In Experiment 3.3, the hydrogen ion concentration was varied whilst the thiosulphate ion concentration was kept constant. The results indicate that hydrogen ions inhibit the increase in the rate at higher hydrogen ion concentrations (see Fig. 3.2). A possible explanation is that the rate expression involves a term in the hydrogen ion concentration in the denominator, e.g.

$$\text{Rate of reaction} = \frac{k'[H^+]}{1 + k''[H^+]} \times [S_2O_3^{2-}] \qquad (1)$$

This can be expressed as:

$$\frac{1}{\text{Rate}} = \frac{1 + k''[\text{H}^+]}{k'[\text{H}^+]} \times \text{Const.}$$

when the thiosulphate ion concentration is kept constant.
Therefore,

$$\frac{1}{\text{Rate}} = \frac{1}{k'[\text{H}^+]} + \frac{k''\,\text{Const.}}{k'} \qquad (2)$$

Since the rate of reaction is proportional to $1/t$, where t is the time taken for the sulphur precipitate to obscure the cross, a graph of t against $1/[H^+]$ will be a straight line if (2), and hence (1), is valid. Such plots are, in fact, found to be linear.

At very high acid concentrations (such as in Question 2 in the exercises at the end of Chapter 3) the rate *falls* with increasing concentration. This may reflect the smaller concentrations of 'free' hydrogen ions in concentrated acid as the ions begin to associate into the ion-pair, H^+Cl^-.

One criticism of the method in this investigation is that the extent of the reaction may be fairly high and therefore the measured rate might not approximate to the initial rate. This is a particularly important consideration using the concentrations given in Experiments 3.2 and 3.3. However, these were chosen specifically because they minimized salt effects (see Appendix B). The extent of reaction followed is appreciably larger than in the clock reactions (Chapter 4), but since this is only an elementary investigation for pre A-level pupils, it need not be mentioned. The experimental results are, in fact, still a useful indication of the relationships between the rate and the concentrations of the reactants.

Appendix B Salt effects

It is a general observation that the rates of reactions in solution are affected by the presence of electrolytes, whether they are reactants, products, or an added 'inert' salt, that is a salt whose ions do not react with either the reactants or the products. These effects are termed *salt effects*.

Reactions between pairs of ions of like charge are usually accelerated by the addition of inert salts, whereas reactions between ions of unlike charge are usually slowed down. The rates of reactions between two uncharged molecules, or between an ion and a molecule are usually only slightly affected by the addition of inert salts and the effect in these cases may be either to increase or to decrease the rate.

The experiments in this book have been designed with the aim of avoiding any complications due to salt effects, and the main purpose of this appendix is to explain how this has been done.

The total electrolyte concentration of a solution is usually measured as its *ionic strength*, I, which is defined as:

$$I = \tfrac{1}{2} \sum m_i z_i^2 \qquad (1)$$

The terms m_i and z_i are the molarity and charge of an ion i, and the summation includes all the ions present in the solution. In general, it is found that salt effects are apparent when the ionic strength is varied: they can be avoided by keeping the ionic strength constant.

Consider the reaction between hydrogen ions and thiosulphate ions (Experiments 3.2 and 3.3). The kinetics were investigated by an initial rate method. In Experiment 3.2 the acid concentration was held constant while the thiosulphate ion concentration was varied. For the mixture with the highest thiosulphate ion concentration, the initial concentrations of the reactants in the reaction mixture were: $Na_2S_2O_3$ 0·08 M and HCl 2 M. The ionic strength of this solution is thus

$$I = \tfrac{1}{2}\left\{\underbrace{(2 \times 0{\cdot}08) \times 1^2 + 0{\cdot}08 \times 2^2}_{Na_2S_2O_3} + \underbrace{2 \times 1^2 + 2 \times 1^2}_{HCl}\right\} \text{mol dm}^{-3}$$

$$= \tfrac{1}{2}(0{\cdot}48 + 4)$$

$$= 2{\cdot}24 \text{ mol dm}^{-3}$$

When the concentration of sodium thiosulphate is halved, the ionic strength becomes

$$I = \tfrac{1}{2}\{2 \times 0{\cdot}04 + 0{\cdot}04 \times 2^2 + 2 + 2\} = 2{\cdot}12 \text{ mol dm}^{-3}$$

Thus halving the initial concentration of sodium thiosulphate causes the ionic strength to decrease by about 5 per cent. This change of ionic strength is small, so that the observed difference between the rates of reaction for the two mixtures may be attributed solely to the effect of changing the thiosulphate ion concentration, with no complications due to any salt effects. This is supported in practice by the plot of the initial rate against the concentration of thiosulphate ions, which is linear within this concentration range.

In Experiment 3.3 the effect on the rate of reaction of varying the concentration of acid was investigated. If the above reaction mixture had been used ($Na_2S_2O_3$ 0·08 M, HCl 2 M), halving the acid concentration would have caused the ionic strength to decrease from 2·24 mol dm^{-3} to 1·24 mol dm^{-3}. Such a change could lead to a considerable salt effect. In order that the ionic strength should remain effectively constant in this series of experiments, different reactant concentrations are needed. The ones used in Experiment 3.3 gave the initial concentrations in the reaction mixture: $Na_2S_2O_3$ 0·25 M, HCl 0·15 M. The initial acid concentration was varied from 0·15 to 0·05 M and calculation shows that this produces a decrease in the ionic strength of about 10 per cent. Thus the considerable curvature found in the plot of $\log_{10}$(rate) against $\log_{10}$(hydrogen ion concentration) cannot be attributed to a salt effect, but must be due to a complex relationship between the rate of reaction and the hydrogen ion concentration (see Appendix A).

Another example of the procedure for avoiding salt effects is provided by the reaction between bromide and bromate ions in acid solution. In Experiment 4.4(*a*) the initial concentrations of the reactants in the reaction mixture were: $KBrO_3$ 0·0014 M, KBr 0·0028 M, H_2SO_4 0·17 M. Hence the ionic strength is 0·514 mol dm^{-3}. If either the potassium bromide or the potassium bromate is diluted to one half of its original concentration, the ionic strength is reduced by less than 0·5 per cent. If, however, the acid concentration were halved, the ionic strength would be reduced by about 50 per cent. The above reaction mixture is therefore suitable for examining the effect on the rate of reaction of varying either the bromide or bromate concentrations, but another reaction mixture is required if the dependence of the rate on the hydrogen ion concentration is to be determined.

In Experiment 4.4(*b*) the initial concentrations of the reactants were: H_2SO_4 0·01 M, KBr 0·4 M, $KBrO_3$ 0·2 M. Calculation shows that halving the acid concentration now causes the ionic strength to decrease by about 2·5 per cent. This mixture is therefore suitable for examining the effect of the hydrogen ion concentration on the rate, but would be unsuitable for determining the effect of varying either the bromide or the bromate concentrations.

These examples illustrate the general rule for avoiding salt effects in *initial rate* experiments: the reactant whose concentration is being varied must have a low concentration compared with the other reactants. As a guide, the maximum variation of the ionic strength should not exceed about 5 per cent. This principle has been used in devising concentrations for Experiments 3.2, 3.3, 4.3, 4.4, and 4.5.

To ignore this principle usually produces non-integral orders, which may differ considerably from the correct order. For example, if the concentrations in Experiment 4.4(*a*) were used to determine the order of the reaction with respect to hydrogen ions, the result would be about 1·6 instead of 2·0.

When a reaction is followed for more than an initial period, the changes of the reactant and product concentrations during the reaction may cause a change in the ionic strength, and hence a salt effect. A simple example is provided by the aqueous hydrolysis of 2-chloro-2-methylpropane (Experiment 9.4.) The initial ionic strength is effectively zero, but during the experiment, as the alkyl halide concentration decreases from about 0·08 M to zero, the hydrochloric acid concentration increases from zero to about 0·08 M, giving a final ionic strength of about 0·08 mol dm^{-3}. Because the reactants in this reaction are neutral molecules, salt effects can be expected to be small, so that it is not surprising that first order kinetics were observed in Experiment 9.4. However, it has been shown† that under different conditions and with a much larger alkyl halide concentration, the first order rate constant can increase by about 30 per cent as the reaction goes to completion, and this has been interpreted as a salt effect.

The avoidance of salt effects in ionic reactions which are followed for a large amount of the reaction is usually more difficult, but there are two general ways of doing it. Consider a reaction between two ionic species A and B.

1. If the concentration of A is made sufficiently larger than that of B, the complete reaction of B will cause only a small change in the ionic strength, and salt effects can be expected to be small. This experiment determines the order with respect to B only, and the procedure has to be repeated with other concentrations if the order with respect to A is to be determined.
2. The alternative procedure is to carry out the reaction in the presence of a high concentration of an 'inert' salt. The ionic strength is then mainly controlled by the concentration of the 'inert' salt which remains constant throughout the experiment. This procedure has not been used in the experiments in this book.

† *J. Chem. Soc.*, 1940, 979.

Appendix C Some theoretical expressions for the bimolecular rate constant, k_2.

Collision Theory

$$k_2 = pYN\exp\left(-\frac{E_c}{RT}\right) \quad \text{[Equation (9), Chapter 7]} \quad (1)$$

For a bimolecular reaction between two gases, the term Y is given by:

$$Y = \sigma_{AB}^2\left[\frac{8\pi RT(M_A + M_B)}{M_A M_B}\right]^{1/2\dagger} \text{m}^3\ \text{s}^{-1} \quad (2)$$

where σ_{AB} is the sum of the radii of the two colliding molecules A and B, R is the gas constant, T is the absolute temperature, and M_A and M_B the respective molecular weights of A and B.

An order of magnitude value for Y can be calculated using reasonable values for each term in (2).

$$\sigma_{AB} \sim 3 \times 10^{-10}\ \text{m}$$

$$8\pi \sim 25$$

$$R = 8{\cdot}3\ \text{kg m}^2\ \text{s}^{-2}\ \text{K}^{-1}\ \text{mol}^{-1}$$

$$T = 500\ \text{K}$$

$$M_A \sim 0{\cdot}1\ \text{kg mol}^{-1}$$

$$M_B \sim 0{\cdot}01\ \text{kg mol}^{-1}$$

Then $Y = 9 \times 10^{-20}\left[25 \times 8{\cdot}3 \times 5 \times 10^2 \times \frac{11 \times 10^{-2}}{10^{-3}}\right]^{1/2}$

$$\sim 3 \times 10^{-16}\ \text{m}^3\ \text{s}^{-1}$$

or $3 \times 10^{-13}\ \text{dm}^3\ \text{s}^{-1}$.

In the calculation of a rate of reaction as in Chapter 7, the units of Y become molecule^{-1} dm^3 s^{-1}.

Hence

$$YN \sim 3 \times 10^{-13} \times 6 \times 10^{23}\ \text{molecule}^{-1}\ \text{dm}^3\ \text{s}^{-1} \times \text{molecule mol}^{-1}$$

$$= 1{\cdot}8 \times 10^{11}\ \text{mol}^{-1}\ \text{dm}^3\ \text{s}^{-1}$$

Transition State Theory

The derivation of the following expression for k_2 is based on the thermodynamics of the activated complex.

$$k_2 = \phi\frac{kT}{h}\exp\left(\frac{\Delta S}{R}\right)\exp\left(-\frac{E_a}{RT}\right) \quad (3)$$

† This expression can be derived from the Kinetic Theory of Gases. For more information the reader should consult a standard reference book on advanced physical chemistry.

where ϕ is an arbitrary fraction between 0 and 1, k is the Boltzmann constant, h is Planck's constant, ΔS is the entropy of activation, R the gas constant, T the absolute temperature, and E_a the activation energy.

It should be noted that a degree of arbitrariness is introduced in (3) by the term ϕ. The terms preceding $\exp(-E_a/RT)$ involve an entropy term because the derivation of (3) includes a consideration of the way in which the atoms of the reactant molecules are rearranged to form the activated complex.

A comparison of the theoretical expressions with the experimental Arrhenius equation

Equations (1) and (3) can be simplified as follows:

Collision Theory $$k_2 = p\ \text{Const.}\ T^{1/2} \exp\left(-\frac{E_c}{RT}\right) \quad (4)$$

Transition State Theory $$k_2 = \phi\ \text{Const.}\ T \exp\left(-\frac{E_a}{RT}\right) \quad (5)$$

Both (4) and (5) closely resemble the Arrhenius equation:

$$k_2 = A \exp\left(-\frac{E}{RT}\right) \quad (6)$$

For any given reaction p and ϕ are constants so that the only difference in form between (4) and (5) and the Arrhenius equation is concerned with the temperature dependence of the pre-exponential terms. In the Arrhenius equation the term A is considered to be independent of temperature. However, the exponential term is extremely dependent on temperature so that the mild temperature dependence of the pre-exponential terms is unimportant.

The bimolecular rate constant for diffusion-controlled reactions

This can be shown to be approximately:

$$k_2 = \frac{8RT}{3\eta}\ \text{mol}^{-1}\ \text{m}^3\ \text{s}^{-1} \quad (7)$$

where η is the viscosity of the solvent. This equation should be compared with (4), (5) and (6) which contain exponential terms: diffusion-controlled reactions have a considerably milder temperature dependence than reactions controlled by the frequency of successful collisions.

An order of magnitude calculation can be made using the following values:

$T = 500$ K

$R = 8{\cdot}3$ kg m^2 s^{-2} K^{-1} mol^{-1}

$\eta \sim 10^{-3}$ kg m^{-1} s^{-1} (typical for most of the common solvents).

$$k_2 = \frac{8 \times 8{\cdot}3 \times 5 \times 10^2}{3 \times 10^{-3}}$$

$$\sim 1{\cdot}1 \times 10^7\ \text{mol}^{-1}\ \text{m}^3\ \text{s}^{-1}$$

or $1{\cdot}1 \times 10^{10}$ mol^{-1} dm^3 s^{-1}

Appendix D Conditions for turbulent flow

The Reynolds number R is defined as:

$$R = \frac{VD\rho}{\eta}$$

where V = the mean velocity of flow of the liquid through the tube (m s^{-1})

D = the diameter of the tube (m)

η = the dynamic viscosity of the liquid (kg m^{-1} s^{-1})

ρ = the density of the liquid (kg m^{-3})

For water at about 20°C, $\eta = 10^{-3}$ kg m^{-1} s^{-1}.

The condition for turbulent flow is that R should exceed about 2000.

Appendix E The definition of rate of reaction

Consider the reaction:

$$\mathrm{aA + bB = pP + qQ}$$

where A,B,P and Q are the reactant and product species and a,b,p and q are their coefficients in the stoichiometric equation. At time t seconds, let there be n_A, n_B, n_P and n_Q molecules of A,B,P and Q present in the reaction mixture. The rate of reaction may be defined as:

$$-\frac{1}{\mathrm{a}}\frac{dn_A}{dt} = -\frac{1}{\mathrm{b}}\frac{dn_B}{dt} = \frac{1}{\mathrm{p}}\frac{dn_P}{dt} = \frac{1}{\mathrm{q}}\frac{dn_Q}{dt} \text{ molecules s}^{-1} \tag{1}$$

The inclusion of 1/a, 1/b, 1/p, and 1/q enables the rate of reaction to be defined unambiguously in terms of any of the species. If N is the Avogadro Number, equation (1) may be rewritten as:

$$-\frac{1}{\mathrm{a}N}\frac{dn_A}{dt} = -\frac{1}{\mathrm{b}N}\frac{dn_B}{dt} = \frac{1}{\mathrm{p}N}\frac{dn_P}{dt} = \frac{1}{\mathrm{q}N}\frac{dn_Q}{dt} \text{ mol s}^{-1} \tag{2}$$

If the volume of the reaction mixture is V dm^3 and is constant, then:

$$-\frac{1}{\mathrm{a}NV}\frac{dn_A}{dt} = -\frac{1}{\mathrm{b}NV}\frac{dn_B}{dt} = \frac{1}{\mathrm{p}NV}\frac{dn_P}{dt} = \frac{1}{\mathrm{q}NV}\frac{dn_Q}{dt} \text{ mol dm}^{-3}\text{ s}^{-1} \tag{3}$$

Representing molar concentrations by c,

$$-\frac{1}{\mathrm{a}}\frac{dc_A}{dt} = -\frac{1}{\mathrm{b}}\frac{dc_B}{dt} = \frac{1}{\mathrm{p}}\frac{dc_P}{dt} = \frac{1}{\mathrm{q}}\frac{dc_Q}{dt} \text{ mol dm}^{-3}\text{ s}^{-1} \tag{4}$$

In the experiment on the decomposition of hydrogen peroxide, the rate of evolution of oxygen was measured, and this is proportional to dn_{O_2}/dt and $1/N \times dn_{O_2}/dt$. [See equations (1), (2).] Similar quantities were measured in the decomposition of benzene diazonium chloride and the thiosulphate-acid reaction. The volume of the reaction mixture was effectively constant in each case.

When samples of the reaction mixture are taken, titres indicate molar concentrations [see equation (4)]. In these cases, the volume of the reaction mixture changes during the experiment, but this is unimportant because concentration is an intensive property of the reaction mixture.

Appendix F The concept of rate determining step

This is more involved than the treatment in the text suggests, and a more detailed account is given here.

Consider an irreversible reaction with the stoichiometry

$$A + 2B \rightarrow \text{products},$$

which proceeds by the two-step mechanism:

$$A + B \underset{k_{-1}}{\overset{k_1}{\rightleftharpoons}} X \qquad \text{Step 1}$$

$$X + B \xrightarrow{k_2} \text{products} \qquad \text{Step 2}$$

Step 1 is reversible, with rate constants k_1 and k_{-1} for the forward and reverse steps respectively, and produces the intermediate X. This intermediate may either decompose to A and B, or it may react to give the products by the irreversible step 2. A rate expression may be derived from this mechanism by using the steady state approximation.

$$\frac{d[X]}{dt} = 0 = k_1[A][B] - k_{-1}[X] - k_2[X][B]$$

Rearranging,

$$[X] = \frac{k_1[A][B]}{k_{-1} + k_2[B]}$$

The rate of reaction is given by:

$$\frac{d[\text{products}]}{dt} = k_2[X][B]$$

Substituting for [X],

$$\text{rate of reaction} = \frac{k_2 k_1[A][B]^2}{k_{-1} + k_2[B]} \qquad (1)$$

Two extreme cases may be examined using equation (1):

(*a*) if $k_{-1} \ll k_2[B]$, then the rate $= k_1[A][B]$;

(*b*) if $k_{-1} \gg k_2[B]$, then the rate $= \frac{k_1}{k_{-1}} . k_2[A][B]^2$.

The steady state approximation assumes that the concentration of X remains effectively constant throughout most of the reaction. This is likely when X is very reactive so that it reacts or decomposes more readily than it is formed: its concentration then remains very low.

Case (*a*) The rate constant, k_{-1}, is so much smaller than k_2[B] that the reverse reaction in step 1 may be ignored. The mechanism may then be written as:

$$A + B \xrightarrow{k_1} X \qquad \text{Step 1 (slow)}$$

$$X + B \xrightarrow{k_2} \text{products} \qquad \text{Step 2 (fast)}$$

X forms slowly but is removed quickly by reaction with B in step 2: its concentration therefore remains very low. The overall rate of reaction is the rate of step 1, the rate determining step.

Case (*b*) This mechanism can be written as:

$$A + B \underset{k_{-1}}{\overset{k_1}{\rightleftharpoons}} X \qquad \text{Step 1 (fast in both directions)}$$

$$X + B \xrightarrow{k_2} \text{products} \qquad \text{Step 2 (slow)}$$

Step 2 is much slower than the reverse of step 1 so that the latter achieves equilibrium with k_1/k_{-1} equal to the equilibrium constant for this step. There is only a slow 'leakage' of X through step 2 so that the concentration of X again remains effectively constant.

This is the type of mechanism proposed for the bromate–bromide-acid reaction and the acid hydrolysis of methyl acetate. However, although step 2 is the slower step, it cannot be described as rate determining: the rate constant is $k_2 k_2/k_{-1}$, and not k_2, so that the overall rate is affected by the values of the rate constants for step 1 as well as by the value of k_2.

Those reactions used in the text which involve this general type of mechanism have all given rate expressions which could be interpreted using the special cases (*a*) and (*b*) above, with the possible exception of the acid-thiosulphate reaction.

Index

Activated complex, 86, 94–6, 105, 127
Activation energy, 86–91, 94
  calculation of, 87
  in catalysed reactions, 103–5
Activation by collision, 138–9
Activators, 114
Active sites, 111, 113
Adsorption in heterogeneous catalysis, 110–12
Arrhenius equation, 86
  plot, 87
  exponential term, 88–90, 96
  pre-exponential term, 89–90
  and theoretical expressions, 92, 162
Arsenious acid–iodide reaction, 149–50
Autocatalysis, 62, 113

Benzene diazonium chloride decomposition, 33–40
  mechanism, 76
'Blue bottle' experiment, 100–2, 115–16
Bromination of acetone, 113
Bromide–bromate reaction, 46–8
  mechanism 77–8, 96
Bromide, oxidation by nitric acid, 61–2
Bromine, dissociation of, 136, 138–9
  association of atoms, 141
Bromomethane hydrolysis, 104

Calcium carbonate–acid reaction, 14
Catalysis
  acid-base, 107
  and activation energy, 103–5
  characteristics of, 22–4
  enzymes, 105, 112–13
  heterogeneous (surface), 105, 107–12
  homogeneous, 105–7
  and intermediates, 98–103
  in the iron(III)–iodide reaction, 106–8
  in redox reactions, 106–7
  in the tartrate–hydrogen peroxide reaction, 102
  in the zinc–acid reaction, 13–14, 157
Catalyst poisons, 114
Chain reactions
  branched, 146

Chain reactions—*cont.*
  chain breakers, 115
  characteristics of, 141–2
  mechanisms, 143–7
  rate expressions, 142
2-Chloro-2-methylpropane hydrolysis, 123
  mechanism, 127–9
Clock reactions,
  theory, 44–6
  bromide–bromate, 46–8
  formaldehyde–hydrogen sulphite, 54
  hydration of carbon dioxide, 54
  hydrogen peroxide–iodide: flowing clock, 63–6; rewound clock, 49–53
  persulphate–iodide, 44–5
Co-enzymes, 114
Collision frequency and rate, 68–9, 82–6
  and vibrational frequency, 139
Collision theory and rate
  gaseous reactions, 68–72, 82–5
  solution reactions, 72–4, 85–6
  defects, 93
  usefulness, 92–3
Collision theory and rate constant, 161
Colorimeter, experiments with, 60–2
Cyclopropane isomerization, 137–9

Deactivation by collision, 139
Decay constant for protoactinium 234, 32–3
Diacetone alcohol depolymerization, 58–9
Dilatometer, 58
Diffusion-controlled reactions, 92–3
  and bimolecular rate constant, 162
Dissociation constants, 59–60

Elementary reactions, 70
Enzymes, 24, 105, 112–13
Equilibrium constant, 149
Ethanol, catalytic decomposition, 109–10
Ethyl acetate saponification, 129–31
  mechanism, 132–3

Fast reactions, 62–7
Fire extinguishers, 147
First-order reaction, 26
Flow methods, 63
Fractional lives, 31
Friedel-Craft reaction, 103

Half-life, 30–3
Harcourt-Esson reaction, 49–52
  mechanism, 74–5, 78–9
Heat of reaction, 95
Hydrogen bromide–oxygen reaction, 70–2
Hydrogen–bromine reaction, 143–5
Hydrogen–halogen reactions, 145–6
Hydrogen–iodine reaction, 146, 150
Hydrogen peroxide
  decomposition by manganese(IV) oxide, 20–2, 54
  decomposition by other catalysts, 22–4
  oxidation of iodide, 49–52, 63–6; mechanism, 74–5, 78–9
  oxidation of tartrate, 102

Induction period, 141–2
Inhibitors, 114–15, 142, 147
Integrated rate expressions, 26–7, 29
Intermediates, 75, 79–80
Iodate–sulphite reaction, 81
Iodide, oxidation by
  arsenious acid, 149–50
  persulphate, 44–5, 54, 116
  hydrogen peroxide, 49–52, 63–6
  iron(III), 106
  nitric acid, 15
Iodination of acetone, 40–4, 60–1
  mechanism, 75–6, 108
Iodoethane hydrolysis, 117–21
  mechanism, 126–7
Ion-pairs, 78, 79, 158
Iron(III) — thionine reaction, 115
Isotopes, use in kinetics, 103, 133

Law of mass action, 154–5
Light, effect on rate, 115
Lindemann-Hinshelwood, 139

Magnesium-hydrochloric acid reaction, 9–13
  —nitric acid reaction, 14–15

Maxwell-Boltzmann distribution, 88–9
Mechanisms
  competing, 78–9
  complex, 70, 80
  elementary, 70, 80
Methyl acetate hydrolysis, 131–2
  mechanism, 133–5
Microscopic reversibility, 153
Molecularity, 72, 78

Nitric oxide–oxygen reaction, 140
Nitrosyl chloride decomposition, 151–4

Order, 26–30
  and molecularity, 72, 78
Orientation effects, 90–2

Potential energy during reaction, 95, 104
Promoters, 114
Pseudo orders, 28

Quantum yield, 141

Radioactive decay, 31–3
Rate, 5–8, 163–4
  by drawing tangents, 7, 21–2
  effect of solvent molecules, 86
  initial, 11–12, 17–18
  measurement, 7–8, 55–6
Rate constant, 26–7
  calculation of, 83, 85, 161, 162
  variation with temperature, 39, 89, 97
Rate determining step, 71, 164–5
Rate expression (equation, law)
  definition, 27
  gaseous reactions, 70
  integrated, 29
  solution reactions, 73
Relaxation techniques, 66–7

Salt effects, 158–60
Second-order rate constants,
  gaseous reactions, 84
  solution reactions, 85
Second-order reaction, 27
Solvent 'cage' effect, 86
Stationary state approximation, 144–5, 164
Sucrose inversion, 56–8

Termolecular steps, 140–1
Thiosulphate–acid reaction, 16–19, 157–8
Third-body restriction, 141
Transition state theory, 93–7, 161–2
Turnover number, 112

Unimolecular steps, 76, 139–40

Velocity constant — *see* Rate constant

Zero order reaction, 28
Zinc–acid reaction, 3, 13–14, 157